DICTIONARY OF ENGINEERING

DICTIONARY OF ENGINEERING

ARUN MITTAL

A P H PUBLISHING CORPORATION
4435-36/7, ANSARI ROAD, DARYA GANJ
NEW DELHI-110 002

Published by
S.B.Nangia
A P H Publishing Corporation
7, Ansari Road, Daryaganj
New Delhi 110002
Ph.: 23274050
E-mail : aphbooks@gmail.com

2026

₹-2995/-

Printed at
Balaji Offset
Navin Shahdara, Delhi 110032

Preface

Engineering is an incredibly broad field which involves the harnessing of mathematical and scientific concepts to create practical and useful things, ranging from automobiles to wind turbines. This field is huge, encompassing people in a wide range of industries, and some people call it "the invisible science," because engineers are often unsung and unheralded, despite the fact that the work they do is very important.

There are a number of subfields encompassed by engineering. Among these are mechanical, electrical, computer, military, civil, environmental, aerospace, and chemical engineering. All of these fields require different types of training; aerospace engineers, for example, learn a great deal about physics and space in the course of their work, while environmental engineers consider issues like pollution control and the impact of humans on their environment. Most engineers pride themselves on being problem solvers. Their field of work involves the analysis of a problem, such as the need for a safe and stable water supply for a city, and the creation of a solution, such as an aqueduct.

Dictionary of Engineering is a vital reference book, offering a wealth of essential information in a convenient, quick-find format. Written in a simple, easily accessible fashion, it containing over 2,000 entries from various brances of engineering such as mechanical, electrical, civil, environmental, aerospace, and chemical. Whether you are a professional, a student, a writer, or a general reader with science curiosity, this comprehensive resource gives you a better understanding of the ideas and concepts you need to know.

Editor

A

A/D Converter : A/D converter is a circuit that converts analog signals into a stream of digital data.

Abiotic Environment : The part of an ecosystem that includes the non-living surroundings.

Abrasive : A hard and wear-resistant material that is used to wear, grind or cut away other material.

ABS Plastic : A type of plastic formed by Acrylonitrile Butacaine Styrene and other chemicals.

Absorption : Dissipation of radio or sound waves as they interact with matter. The absorbing of light waves without reflection or refraction.

Absorption Wavemeter : An instrument used to measure audio frequencies.

Absorption, Law of : In Boolean algebra, the law which states that the odd term will be absorbed when a term is combined by logical multiplication with the logical sum of that term and another term, or when a term is combined by logical addition with the logical product of one term and another term.

AC (Alternating Current) : A signal or power source that varies with time, switching polarities. Typically, sinusodial and constant frequency.

AC Coupling : The interconnection between two circuits through a capacitor or other device which passes AC voltage but blocks any DC component. In the circuit below C3 blocks the DC component of Q1 from effecting the operation of Q2.

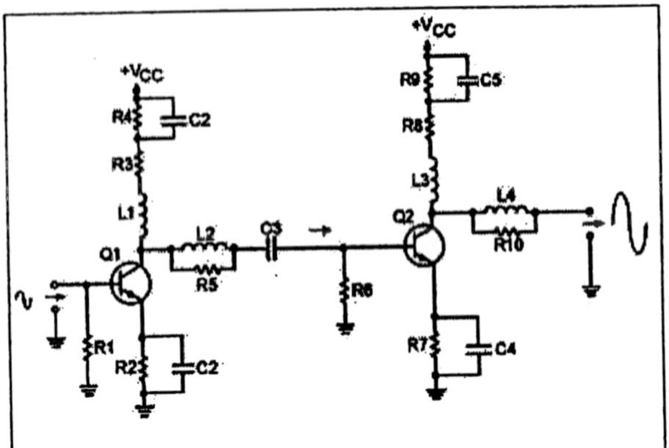

AC Coupled Transistor Amplifier

AC/CDI : A type of Capacitor Discharge ignition system which uses a special alternator to provide both a high voltage to charge up a capacitor and a timed trigger pulse to discharge it through the coil. Mostly used on motorcycles and utility vehicles having no battery.

Acacia : A durable wood used for palings and posts. A lot of the gums in common use are obtained from varieties of the Acacia.

Accelerating Anode : An electrode charged several thousand volts positive and used to accelerate electrons toward the front of a cathode-ray tube.

Acceleration Enrichment : Extra fuel provided when the throttle is opened. May be in the form of extra injection pulses or enrichment of the basic fuel law, often arranged to decay gradually over two or three seconds. Usually variable according to rate at which the throttle is opened.

Acceleration Servosystem : A servo-system that controls the acceleration (rate of change in velocity) of a load.

Accelerometer **:** A device that measures the acceleration to which it is subjected and develops a signal proportional to it.

Acceptor Impurity **:** An impurity which, when added to a semiconductor, accepts one electron from a neighbouring atom and creates a hole in the lattice structure of the crystal.

Access Bus **:** Is a low speed 4-wire serial bus once used on Personal Computers. Access. Bus uses the I2C [IIC] bus as the electrical hardware interface. The Access bus has a maximum speed is 100kbps over a maximum cable distance of 10 meters, however a repeater may be used.

Access Time **:** The difference in time between when data is requested and when data is delivered. The time lapsed between a given command and when the function is performed.

Accessory **:** An assembly of a group of parts or a unit which is not always required for the operation of a set or unit as originally designed but serves to extend the functions or capabilities of the set, such as headphones for a radio set supplied with a loudspeaker, a separate power unit for use with a set having a built-in power supply, or a remote control unit for use with a set having integral controls.

Acme Thread **:** A screw thread having a 29 degree included angle. Used largely for feed and adjusting screws on machine tools.

Acorn Tube **:** A very small tube with closely spaced electrodes and no base. The tube is connected to its circuits by short wire pins that are sealed in a glass or ceramic envelope. The acorn tube is used in low power uhf circuits.

Acoustic Noise **:** The unintended sound a component produces while operating. Some fields apply this to mean an audible sound, while other fields take it to mean any sound.

ACPI (Advanced Configuration and Power Interface) **:** An industry-standard specification (co-developed by Hewlett-Packard, Intel, Microsoft, Phoenix, and Toshiba, for operating-system-directed power management for laptop, desktop, and server computers.

Activated Sludge : An active population of microorganisms used to treat wastewater, or the process in which the organisms are employed.

Active High : Refers to a non-inverted output, the output is true when high and false when low. When the Sum-of-Products expression is true the output is high.

Active Low : Refers to an inverted output, the output is true when low and false when high. When the Sum-of-Products expression is true the output is low.

Active Satellite : A satellite that amplifies the received signal and re-transmits it back to earth.

Active Serial Interface : A term used with programming an FPGA. In the Active Serial AS configuration scheme, Altera FPGAs are configured with serial configuration devices. Also called Active Serial Mode, or Active Serial Scheme; where Active Serial may be Abbreviated as AS.

Actuating Force : The force applied to the actuator to operate the contacts.

Actuator : The part of a switch that is acted upon to cause the switch to change contact connections; for example, toggle, pushbutton, and rocker.

Acute Angle : An angle that is less than 90 degrees.

Adapter : A component that translates from one type of fixture to another. As in a connector that translates from an SMA to BNC connector.

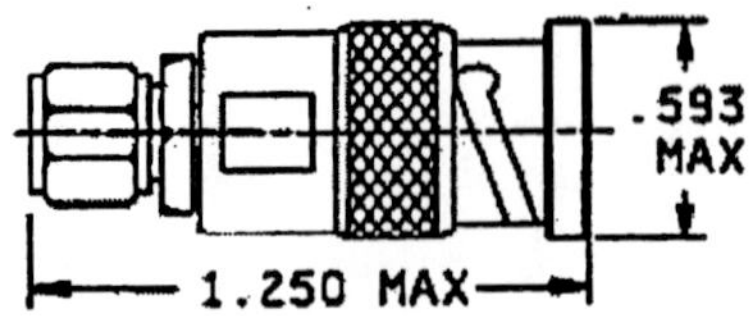

BNC Adapter

Adaptive Memory : A capability provided in more advanced systems to 'learn' the range of signals encountered from the various sensors during normal use and make corrections for tolerance and adjustment errors.

Add/Drop Multiplexer (ADM) : A synchronous transmission network (SDH or Sonet) can carry multiple channels. An Add/Drop Multiplexer is a device that adds (inserts) or drops (removes) lower-data-rate channel traffic from the higher-rate aggregated channel.

Addendum : That portion of a gear tooth that extends from the pitch circle to the outside diameter.

Adhesive : A substance that bonds together the surfaces of two other materials.

Adhesive Terminology : A list of terms used in the description of adhesive properties and manufacture.

ADM (Add/Drop Multiplexer) : A synchronous transmission network (SDH or Sonet) can carry multiple channels. An Add/Drop Multiplexer is a device that adds (inserts) or drops (removes) lower-data-rate channel traffic from the higher-rate aggregated channel.

ADPCM (Adaptive Differential Pulse Code Modulation) : A compression technique that encodes only the difference between sequential samples.

ADSL (Asymmetric Digital Subscriber Line) : A method for moving data over regular phone lines. An ADSL circuit carries much more data than a modem can encode on a regular phone connection. ADSL rides on the regular phone wires coming into the subscriber's premises (twisted pair copper).

Adsorption : A surface phenomena in which a solute (soluble material) concentrates or collects at a surface (the adsorbent).

Advanced Wastewater Treatment : The removal of any dissolved or suspended contaminants beyond secondary treatment, often this is the removal of the nutrients nitrogen and/or phosphorus.

AEC-Q100 : A qualification test sequence for integrated circuits developed by the AIAG automotive organization.

Aeration : Intimate contact of the atmosphere and water to add air (oxygen) to the water. The term is also applied to gas stripping where an undesirable gas is removed from the water.

Aerobes : Organisms which require molecular oxygen as an electron acceptor for energy production.

Aerobic Process : A process which requires molecular oxygen.

AES/EBU (Audio Engineering Society/European Broadcasting Union) : AES/EBU is one of the common names for a digital audio transfer standard. The standard is also known as XLR because the AES/EBU digital interface is usually implemented using 3-Pin XLR connectors, which happens to be the same type connector used in a professional microphone. One cable carries both left and right-channel audio data. AES/EBU is an alternative to the S/PDIF Sony/ Philips Digital Interface standard. S/PDIF is a an audio transfer file format which uses an RCA connector.

AFE (Analog Front End) : The analog portion of a circuit which precedes A/D conversion.

After-Start Enrichment : Extra fuel provided to prevent stall out after the starter has been disengaged, decaying away over 30 seconds or so.

AGC (Automatic Gain Control) : A circuit that modulates an amplifier's gain, in response to the relative strength of the input signal, in order to maintain the output power.

AGP Bus : The Accelerated Graphics Port bus was widely used as a video expansion bus on personal computers from 1997 to 2006. The AGP electrical standard was derived from the Parallel PCI bus. The fastest speed; AGP 8x used a 533MHz clock with 32 bytes/clock transfers, with a Bandwidth of 2.1GBps. The AGP Bus was rendered obsolete by the introduction of the PCIe interface.

Ah (Ampere-hours) : A measure of battery capacity. A 4Ah battery could, for instance, deliver 1A for 4 hours, 1/2A for 8 hours, etc.

Air-control Panel : Panel that monitors the dry-air input at each user equipment.

Air-core Inductor : An inductor composed of two or more coils that are wound around a non-metallic core.

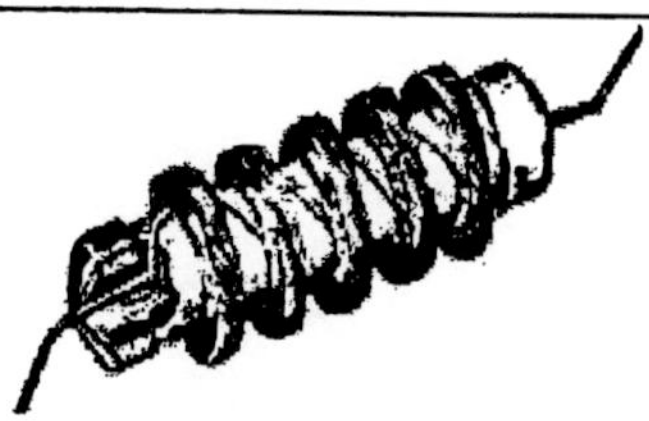

Inductor

Air-core Transformer : A transformer composed of two or more coils that are wound around a non-metallic core.

Air Discharge : A method for testing ESD-protection structures in which the ESD generator is discharged through an air gap between the generator and the device under test (DUT).

Air Injection : The introduction of air into the exhaust stream to promote oxidising reactions of CO and HC pollutants, usually, but not always, in combination with a catalyst. Usually directed into the exhaust port or manifold from an air pump but some early systems relied on low pressure pulses in the exhaust manifold at light throttle to draw air through a one way valve. The main application is from a start up, particularly from cold, when the extra air helps to promote reactions aiding catalyst light-off. The system then switches off.

Air Temperature Sensor : A two terminal device, situated in the induction tract, the resistance of which changes with temperature. The resistance normally falls with rising temperature hence these devices are sometimes correctly described as 'negative temperature coefficient (NTC.) thermistor'.

Airflow Meter : A device which measures the amount of air passing through the induction tract into the engine. Ideally air mass flow is measured rather than volume which requires additional correction for temperature and for barometric pressure.

Airflow Meter

Airflow Meter-Cone Type : A delightfully simple device composed of a cone, suspended in a profiled air tract and pointing into the air stream, which moves laterally against spring pressure by an amount related to the airflow around it. The cone moves the slide of a linear potentiometer to produce the voltage output signal.

Airflow Meter-Flap Type : A common type of airflow meter consisting of a spring loaded flap which hinges within a specially profiled air channel. A complex potentiometer converts the angular displacement of the flap, which varies according to the volume of air passing through the channel, into a voltage signal. An air temperature sensor is usually built into the device but barometric correction is not catered for.

Airflow Meter -Hot Film Type : A variation on the hot wire theme in which a metal foil film is heated to maintain a constant temperature relative to and above that of the passing air as sensed by a thermistor. The load signal is derived from the current required to maintain the film temperature, this current being a measure of mass air flow through the device.

Airflow Meter-Hot Wire Type : A widely used type of airflow meter in which the passage of air has a cooling effect on a heated wire which is part of a balanced resistor circuit, another leg of which applies correction for air temperature. An amplifier maintains the circuit in equilibrium and produces a signal voltage which rises in proportion to mass air flow according to the cooling effect on the wire. In some versions the output is a continuous train of pulses, the frequency of which is modulated by the hot wire voltage.

Airflow Meter-Vortex Type : A device which induces a vortex, or helical flow pattern, in the air flowing through a passage and by measuring the frequency of the vortex is able to make an accurate measurement of volumetric flow. Correction for temperature and barometric pressure is still necessary to arrive at the true mass flow.

AISG (Antenna Interface Standards Group) : AISG creates open specifications for antenna-line control and monitoring for 3G systems.

A-Law : A companding algorithm used in digital communications systems.

Alcohol : An organic compound with one or more hydroxyl "-OH" groups.

Aldehyde : An organic compound with a carbonyl at one end of a hydrocarbon group.

Alder : This wood is very durable in water and not so durable in dry conditions. Historically it was used a lot in pumps, piles and sluices.

Aliasing : In A/D conversion, the Nyquist principle states that the sampling rate must be at least twice the maximum bandwidth of the analog signal. If the sampling rate is insufficient, then higher-frequency components are "undersampled" and appear shifted to lower-frequencies. These frequency-shifted components are called aliases. The frequencies that shift are sometimes called "folded" frequencies because a spectral plot looks like it was folded to superimpose the higher frequency components over the sub-Nyquist portion of the band.

Allowance **:** The prescribed difference in dimensions of mating parts to provide a certain class of fit.

Alloy **:** A metal made from a combination of other metals or non-metals to produce certain desirable quantities.

Alloyed Junction **:** A junction formed by re-crystallization of a molten region of P-type material on an N-type substrate, or vice versa.

Alpax **:** Trade name for material used mainly for castings. Low density and corrosion resistance. 87 % Aluminium 13 % Silicon

Alpha **:** The emitter-to-collector current gain in a common-base circuit of a transistor.

Alternating Copolymer **:** A polymer, composed of two different repeating mere, in which the different mer units systematically alternate positions along the molecular chain.

Altitude **:** The vertical distance of an aircraft or object above a given reference, such as ground or sea level. Additional reference information may be obtained via Airborne Environment Classification per MIL-HDBK-217.

Aluminium **:** Aluminium is the basis for a number of alloys.

Aluminum Creep **:** The movement of aluminum wire from a point where pressure is applied. The "retreat" of heated aluminum wire as it cools.

AM (Amplitude Modulation) **:** A modulation method in which the carrier amplitude changes with the input signal amplitude.

Ambient Temperature **:** Temperature of the air surrounding a component.

Ambient Temperature Sensor **:** Temperature sensor used to measure the temperature of the air that surrounds a component (the ambient temperature).

Amino Acid **:** A functional group which consists of a carbon with a carboxylic acid, "-COOH" and an amine, "-NH_2." These compounds are the building blocks for proteins.

Ammonium **:** The ion NH_4 behaves like a mono valent alkali metal.

Amorphography : The branch of science concerned with the determination of amorphous solid structures and their systematic classification.

Ampacity : The amount of current a conductor can carry without exceeding its specified temperature, in amperes.

Ampere : Ampere(s), the unit of electrical current. Current is defined as the amount of charge that flows past a give point, per unit of time. The symbol I is used for current in equations and A is the abbreviation for ampere.

Ampere-hour : A measure of charge (or current flow over time).· : One ampere-hour (or amp-hour or Ah) is a current of one ampere flowing for one hour. The amount of charge transferred in that hour is 3,600 coulombs (ampere-seconds). A milliampere-hour (mAh or milliamp-hour) is a thousandth of an amp-hour. An ampere-second (As or amp-second) is an amp supplied for one second. A common use of the term is rating energy storage device capacity, especially rechargeable batteries. For example, a 12-volt, 7Ah rechargeable battery used in an alarm system will supply an amp at the rated voltage range for seven hours, 2 amps for 3.5 hours, etc. If my alarm consumes 250mA, this battery would operate the system for 28 hours.

Amplifier : An electrical circuit that produces an output that is a replica of the input. The output may be scaled or have increased drive, or it may provide isolation (so changes in output conditions do not affect the input or other outputs. It may perform other transformations (e.g. filtering or logarithmic drive).

Amplifier Class : Amplifier circuit types are divided into "classes" which describe whether the amplifier operates in a linear or switching mode, and any techniques used to restore linearity of output.

AMPS (Advanced Mobile Phone System) : An analog only, 1G standard that operates in the 800MHz to 900MHz frequency band. It is still widely used in the United States.

AMR (Automatic Meter Reading) : A system installed to read a utility meter remotely.

Anabolism : Biosynthesis, the production of new cellular materials from other organic or inorganic chemicals.

Anaerobes : A group of organisms that do not require molecular oxygen. These organisms, as well as all known life forms, require oxygen. These organisms obtain their oxygen from inorganic ions such as nitrate or sulfate or from protein.

Anaerobic Process : A process which only occurs in the absence of molecular oxygen.

Analog : A system in which an electrical value (usually voltage or current, but sometimes frequency, phase, etc.) represents something in the physical world. The electrical signal can then be processed, transmitted, amplified, and finally, transformed back into a physical quality. For example: a microphone produces a current that is proportional to sound pressure. Various stages amplify, process, modulate, etc. Ultimately, a varying voltage is presented to a speaker which converts it back to sound waves. By contrast, a digital system handles a signal as a stream of numbers.

Analog Switch : An analog switch (sometimes just called a "switch") is a switching device capable of switching or routing analog signals, based on the level of a digital control signal. Commonly implemented using a "transmission gate," an analog switch performs a function similar to that of a relay. For example, an analog switch can turn an audio signal on or off based on a MUTE signal; or analog switches could send one of two signals to a head phone amplifier. Most commonly implemented using CMOS technology integrated circuits.

Analog Temperature Sensor : Temperature sensor with a continuous analog voltage or current output that is related, usually linearly, to the measured temperature.

Analogue to Digital Converter (ADC). A type of microchip which converts a voltage input signal to a binary digital code. Multi-channel ADCs are commonly used to convert sensor information into a form which a microprocessor can make use of. Sometimes the microprocessor itself contains the ADC.

AND : Combining two signals so that the output is on if both signals are present. This can be accomplished by an AND logic gate (two inputs, one output which is high if both inputs are).

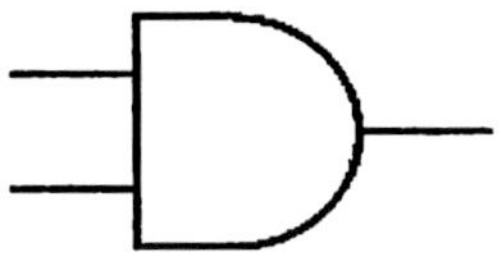

AND Gate Symbol

Angle Iron : An iron or steel structural member that has been cast, rolled, or bent (folded) so that its cross section is L-shaped.

Angle Plate : A precision holding fixture made of cast iron. Steel, or granite. The two principal faces are at right angles and may be slotted for holding the work or clamping to a table.

Angles : The inclination of one line to another. One complete revolution equals 360° (degrees) or 2p radians.

Angstrom : Angstrom is an internationally recognized unit of length equal to 0.1. (10-10 meter). The unit used to define the wavelength of light waves.

Anisotropic : Exhibiting different values of a property in different crystallo graphic directions.

Anneal : To soften and relieve strains in any solid material, such as metal or glass, by heating to just below its melting point and then slowly cooling it. This also generally lowers the tensile strength of the material, while improving its flex life.

Annular Ring : The portion of a conductive material that completely surrounds a hole.

Anode : A positive electrode of an electrochemical device (such as a primary or secondary electric cell) toward which the negative ions are drawn. The semiconductor-diode terminal that is positive with respect to the other terminal when the diode is biased in the forward direction.

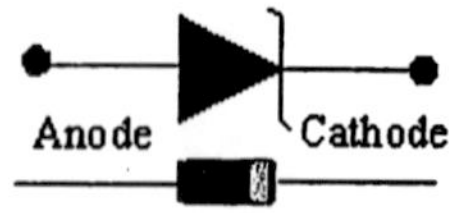

Diode

Anoxic process : A process which occurs only at very low levels of molecular oxygen or in the absence of molecular oxygen.

Antenna : A conductor or set of conductors used to radiate RF energy into space or to collect RF energy from space or to do both.

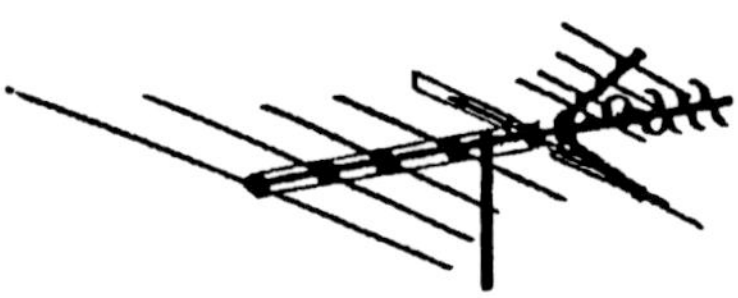

Antenna

Antenna Beam Width : Width of a radar beam measured between half-power points.

Antenna Coupler : A device used for impedance matching between an antenna and a transmitter or receiver.

Antenna System : Routes RF energy from the transmitter, radiates the energy into space, receives echoes, and routes the echoes to the receiver

Anthropogenic : Of, made, or caused by human activity or actions.

Anti-Aliasing : An anti-aliasing filter is used before A/D conversion. It is a lowpass filter that removes signal components above the Nyquist frequency, thereby eliminating their sampled replicas (aliases) in the base band.

Antiferromagnetism : A phenomenon observed in some materials in which complete magnetic moment cancellation occurs as a result of antiparallel coupling of adjacent atoms or ions. The macroscopic solid possesses no net magnetic moment.

Antipads : are clearances to provide separation between vias and voltage planes to avoid shorting the signals to the voltage rails as the via traverses the Printed Wiring Board. The diameter of the Antipad should be minimized to reduce EMI effects, but maximized to decrease shorting and voltage breakdown of adjacent signals.

Antiseize Compound : A silicon-based, high-temperature lubricant applied to threaded components to aid in their removal after they have been subjected to rapid heating and cooling.

Antitransmit-Receive Tube (ATR) : A tube that isolates the transmitter from the antenna and receiver. Used in conjunction with ATR tube.

Anvil : A heavy iron or steel block upon which metal is forged or hammered-. Also the fixed jaw on a micrometer against which parts are measured.

APC *(Automatic Power Control)* : Feature in laser drivers that uses feedback from the laser to adjust the drive, to keep the laser's output constant.

APD *(Avalanche Photo Diode)* : A photo diode designed to take advantage of avalanche multiplication of photo current to provide gain. As the reverse-bias voltage approaches the breakdown voltage, hole-electron pairs created by absorbed photons acquire sufficient energy to create additional hole-electron pairs when they collide with ions. Thus a multiplication or signal gain is achieved.

API *(Application program Interface)* : A software layer that allows a system to be programmed via a defined set of commands.

APM *(Advanced Power Management)* : Power management standard for computers that provides five power states.

APON *(ATM (-based) passive optical network)* : Applies to mechanical systems (e.g. that use relays or manual switches) and to solid-state analog multiplexers and switches.

Apparent Drift: The effect of the earth's rotation on a gyro that causes the spinning axis to appear to make one complete rotation in one day.

Apparent Power: That power apparently available for use in an ac circuit containing a reactive element. It is the product of effective voltage times effective current expressed in volt-amperes. It must be multiplied by the power factor to obtain true power available.

Apple Desktop Bus (ADB) : The ADB or Apple Desktop Bus was a serial bus used by Apple computer to drive the mouse and keyboard interfaces. The cable consisted of one data line ADB, a power line +5v and a ground line. The maximum data rate was 125kbps, actual bus speed was much lower. The ADB bus was rendered obsolete by the introduction of the Firewire interface on Apple Computer products.

APQP (Advanced Product Quality Planning) : System developed by the AIAG automotive organisation to communicate common product quality planning and control plan guidelines for suppliers to the automotive industry.

Apron : That portion of a lathe carriage that contains the clutches, gears, and levers for moving the carriage. It also protects the mechanism.

Arbitration: The process of selecting one communication node on a bus over another.

Arbor Press : A hand-operated machine tool designed for applying high pressure for the purpose of pressing together or removing parts. A shaft or spindle for holding cutting tools; most usually on a milling machine.

Architecture: Defines how a system is connected or interconnected and in some cases how the components are interconnected.

Armature Losses: Copper losses, eddy current losses, and hysteresis losses that act to decrease the efficiency of armatures.

Armature Reaction: The effect in a dc generator of current in the armature creating a magnetic field that distorts the main field and causes a shift in the neutral plane.

Armature: In a relay, the movable portion of the relay. The windings in which the output voltage is generated in a generator or in which input current creates a magnetic field that interacts with the main field in a motor.

Armor: A flexible metal 'shield' that protects a component from damage.

Armstrong Oscillator: A type of oscillator. The Armstrong Oscillator is used to produce a sine-wave output of constant amplitude and of fairly constant frequency within the rf range.

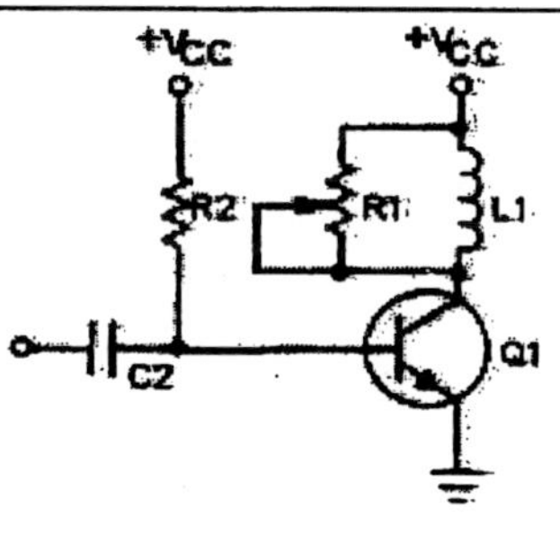

Armstrong Oscillator

Aromatic : A form of bonding in which ring compounds share electrons over more than two atoms. The electrons are de-localised. This leads to unusual ring stability.

Array: A group of devices or components. A number of devices contained within a single package, but not connected to each other.

Arrest Point : When heating or cooling a metal or alloy slowly the temperature may cease to rise or fall respectively due to heat absorption or evolution at the material undergoes a structural change.

Artificial Transmission Line: An LC network that is designed to simulate characteristics of a transmission line.

ASCII: The American Standard Code for Information Interchange. A common method used by computers to code alphabetic letters into a binary code.

Ash **:** A very tough and elastic timber used for oars, blocks and pulleys, spokes of wheels and milk-pails. However, it is not very flexible and is difficult to work. Ash bark is used tanning and the ashes of ash is used for potash.

Aspect Ratio: The ratio between the width of an object and the height of the object.

Assembly: A number of parts or subassemblies, or any combination thereof, joined together to perform a specific function. As in a Printed Wiring Board Assembly, which comprises the Printed Wiring Board [PWB] and all the components attached to the PWB.

Astable Multivibrator: A multivibrator that has no stable state. Also called free-running because it alternates between two different output voltage levels during the time it is on. The frequency is determined by the RC time constant of the coupling circuit.

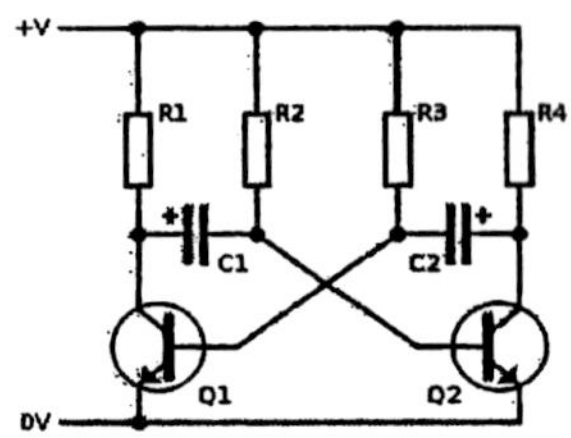

Astable Multivibrator

Astigmatism: An aberration of a lens with spherical surfaces such that the image of a point not lying on the optical axis is a pair of short lines normal to each other and at slightly different distances from the lens. Radial and tangential lines are in focus in different image planes.

Asymmetric Digital Subscriber Line (ADSL) **:** A method for moving data over regular phone lines. An ADSL circuit carries much more data than a modem can encode on a regular phone connection. ADSL rides on the regular phone wires coming into the subscriber's premises.

Asymmetrical Multivibrator: A multivibrator that generates rectangular waves.

Asynchronous Communication: Communication in which no clocking information is passed between stations.

Asynchronous Logic: A Digital circuit whose output only depends on the combination logic gates used and the propagation delay through those gates. A logic circuit that does not use a Flip Flop or other synchronizing element.

Asynchronous Transmission: Parallel Information is synchronized individually [per transfer] by hand-shaking as in DAV [Data Available] and DAC [Data Accepted], or REQ and ACK lines. Serial Information is synchronized individually [per block] via 'START' and 'STOP' bits. Information is passed between nodes with out a separate clock signal to indicate when data on the transmission line is valid.

Atactic : A type of polymer chain configuration wherein side groups are randomly poitioned on one side of the polymer backbone or the other.

Attached Growth Reactor : A reactor in which the microorganisms are attached to engineered surfaces within the reactor. Examples of attached growth reactors are the trickling filter and the rotating biological contactor.

Attenuation : The decrease in magnitude of a signal as it travels through any transmitting medium, such as a cable or circuitry. The ability of a filter circuit to reduce the amplitude of unwanted frequencies to a level below that of the desired output frequency.

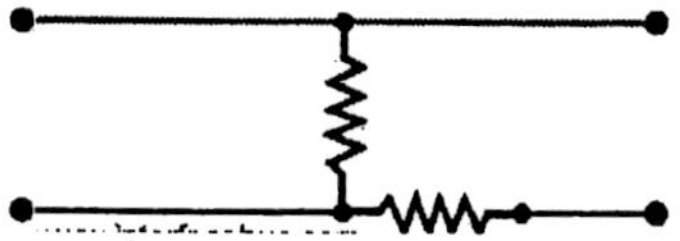

L-Pad Attenuator Circuit

Attenuation Coefficient : The factor relating transmission line attenuation to unit distance.

Attenuation Unbalance : A measure of the inequality of attenuation between two pairs in a multi-pair cable.

Attenuator Pad : A component inserted into a transmission path to introduce transmission loss.

Attraction : The force that tends to make two objects approach each other. Attraction exists between two unlike magnetic poles (north and south) or between two unlike static charges.

Audio : Sound between the range of 15 hertz (15 Hz) and 20 kilohertz (20 kHz).

Audio Amplifier : An amplifier designed to amplify frequencies between 15 hertz (15 Hz) and 20 kilohertz (20 kHz). The transistor Q1 is the amplifier in the circuit to the right.

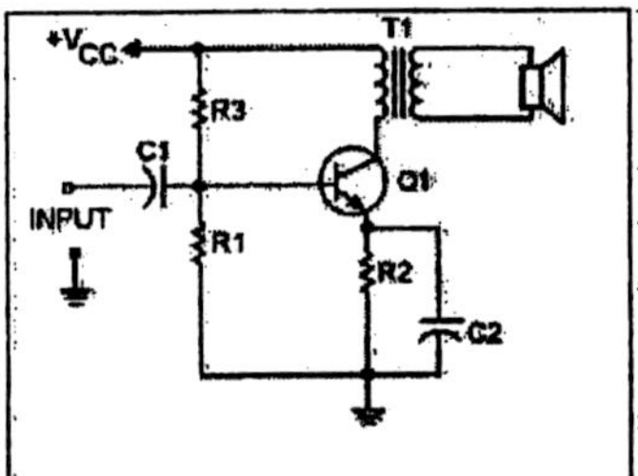

Audio Amplifier

Audio-Frequency-Tone Shift. A system that uses amplitude modulation to change DC mark and space impulses into audio impulses.

Austenite : Face-centered cubic iron; also iron and steel alloys that have the FCC structure.

Auto Shutdown : A feature in EIA-232 interface devices which puts the IC into a low-power shutdown mode when no signal is present on the EIA-232 bus.

Automatic Gain Control : A circuit used to limit variations in the output signal strength of a receiver.

Automatic Stop : A device which may be attached to any of several parts of a machine tool to stop the operation of the machine at any predetermined point.

Autoshutdown Plus : A feature in EIA-232 interface devices which puts the IC into a low-power shutdown mode when no signal is present on the bus or the transmitter inputs.

Autotransformer : An autotransformer is a transformer that uses a common winding for both the primary and secondary windings. Essentially an inductor with a center-tap, an autotransformer is often used in power-supply boost-converter applications to achieve a higher output voltage, while limiting the peak flyback voltage seen by the power switch.

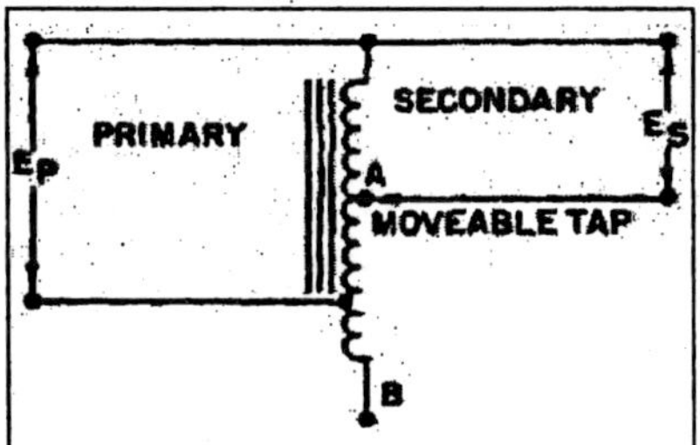

Autotransformer

Autotrophic : Organisms which utilise inorganic carbon for synthesis of protoplasm. Ecologists narrow the definition further by requiring that autotrophs obtain their energy from the sun. In microbiologist parlance, this would be a photoautotroph.

Auxiliary Air Valve : A valve which allows the admittance of extra air for idling, controlled according to engine temperature and possibly by other parameters

Avalanche Effect : A reverse breakdown effect in diodes that occurs at reverse voltages beyond 5 volts. The released electrons are accelerated by the electric field, which results in a release of more electrons in a chain or "avalanche" effect.

Average Power : The peak power value averaged over the pulse-repetition time. Output power of a transmitter as measured from the start of one pulse to the start of the next pulse.

Average Seek Time : For Hard Disk Drives [HDD] the average seek time is the amount of time it takes for a head to seek out the specified address, which is calculated over a large number of random seeks.

Average Value (of AC) : The average of all the instantaneous values of one-half cycle of alternating current.

A-Weighting : A-weighting is a standard weighting curve applied to audio measurements, designed to reflect the response of the human ear. Sound-pressure levels derived using A-weighting are denoted by "dBA," or A-weighted dB levels.

AWG (American Wire Gauge) : A measure of wire thickness (which also dictates cross-sectional area, and for a given material, ampacity). Example: 24 AWG wire has a nominal diameter of 0.0201 inch or 0.511 mm. Also called the Brown and Sharpe Wire Gauge. Note that steel wire is measured by a different gauge. AWG only applies to wire used to conduct electricity.

Axial Lead : An electrical component that has its terminals [leads] protruding out two sides of the component body, along the devices axis. A through-hole component.

Axis : The line, real or imaginary, passing through the center of an object about which it could rotate; a point of reference *or* the axis of a graph.

Azimuth : Angular measurement in the horizontal plane in a clockwise direction.

B

B (Bel) : Measurement of a signal's power compared to a reference; also, measurement of sound pressure.

BMEP (Brake Mean Effective Pressure) : The mean pressure acting on the piston crown of an engine as calculated from the torque measured on a dynamometer thus taking into account all internal losses. Equal to specific torque (lbs/ft per litre) x 2.464. A good modern production engine would produce a figure of around 175 p.s.i. F3000 racing engines, with r.p.m. restricted by rules, achieve about 250 p.s.i. but higher revving F1 engines tend to be around 200-210 p.s.i.

Babbit's Metal : Patented by Isaac Babbitt as a bearing metal due to it's excellent low friction properties.

Back Gears : Gears fitted to a machine to increase the number of spindle speeds obtainable with a cone or step pulley belt drive.

Back Rake : The angular surface ground back from the cutting edge of cutting tools. On lathe cutting tools. The rake is positive if the face slopes down from the cutting edge toward the shank, and negative if the face slopes upward toward the shank.

Back Resistance : The larger resistance value observed when you are checking the resistance of a semiconductor.

Back Shell : The rear portion of a connector.

Background Noise : The total system noise in the absence of a signal.

Backlash : The lost motion or looseness (play) between the faces of meshing gears or threads.

Backup Step-Up : Step-up, switching-regulator power supply with a backup battery switch over.

Bad Block/Bad Sector : An area on a hard disk drive that is damaged or detected as reliable.

Baghouse Filter : A fabric filter device used to remove particulate air pollutants.

Bainite : A Fe-C composition consisting of a fine dispersion of cementite in alpha-ferrite. It is an austenitic transformation product that forms at temperatures between those at which pearlite and martensite transformations occur.

Balanced Code : In PCM systems, a code constructed so that the frequency spectrum resulting from the transmission of any code word has no DC component. A code that has a finite digital sum variation PCM.

Balanced Line : A cable or circuit having two identical conductors with the same electromagnetic characteristics in relation to other conductors and to ground, as in a pair of conductors.

Balanced Mixer : A circuit or device that generates the sum and difference frequencies of two input signals.

Balanced Pair : A transmission line in which the two conductors are electrically identical and symmetrical with respect to a common reference point, usually earth.

Balanced Phase Detector : A circuit that controls the oscillator frequency.

Ballast or Power Resistor : A resistor or resistor pack used to limit current flow through injectors or, in some early examples, the ignition coil, dissipating energy as heat.

Balun : A device used to couple a balanced system to an unbalanced system.

Banana Jack : A type of connector. The Banana Jack may be a male on one end and a female on the other end, or a completely different style connector.

BNC to Banana Jack Connector

Band Gap Energy **:** For semiconductors and insulators, the energies that lie between the valence and conduction bands.

Bandpass Filter **:** A filter that allows a narrow band of frequencies to pass through the circuit. Rejects or attenuates frequencies that are either higher or lower than the desired band of frequencies. A filter that ideally passes all frequencies between two non-zero finite limits and bars all frequencies not within the limits. The cutoff frequencies are normally taken to be the 3-dB points.

Band-Reject Filter **:** A tuned circuit that does not pass a specified band of frequencies.

Bandsaw **:** A power saw. The blade of which is a continuous, narrow, steel band having teeth on one edge and passing over two large pulley wheels.

Bandwidth **:** The width of a communication channel, measured as frequency. The difference between the highest usable frequency of a device (upper frequency limit) and the lowest usable frequency of the device (lower frequency limit) - measured at the half-power points. The difference between the limiting frequencies within which performance of a device, in respect to some characteristic, falls within specified limits. The difference between the limiting frequencies of a continuous frequency band. The capacity of a telecom line to carry signals. The necessary bandwidth is the amount of spectrum required to transmit the signal without distortion or loss of information.

Bar Stock : Metal bars of various lengths, made in flat, hexagon, octagon, round, and square shapes from which parts are machined.

Barretter : A type of bolometer characterised by an increase in resistance as the dissipated power rises.

Base : A reference value. A number that is multiplied by itself as many times as indicated by an exponent. Same as radix. The region between the emitter and collector of a transistor that receives minority carriers injected from the emitter. The element in a transistor that controls the flow of current carriers. The element that corresponds to the control grid of an electron tube.

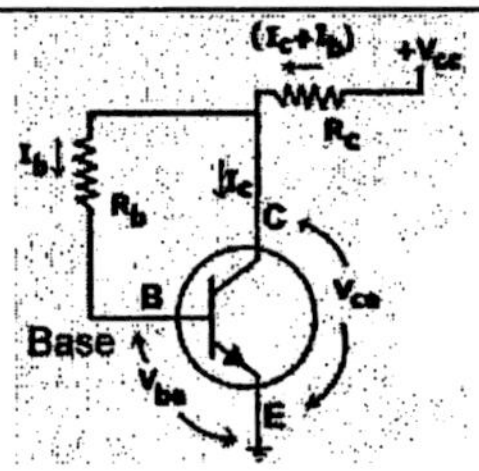

Transistor Base

Base Station : A base station (or base station) is a wireless transceiver at a fixed location (e.g. atop a telephone pole) which is part of a wireless communications network, e.g. the cell phone network. Typically, the base station connects to any cell phones in its area and relays the calls to the wired network. A femto base station is a smaller, personal base station which might cover a home or building and connect via a DSL Internet connection.

Base-Injection Modulator : Similar to a control-grid modulator. The gain of a transistor is varied by changing the bias on its base.

Baseline : The electrical signal from a sensor when no measured variable is present. Often referred to the output at no-load condition.

Basic : Beginner's All-purpose Symbolic Instruction Code, a non-graphical programming language. Also refer to a short example of Hp Basiccable. An assembly of one or more insulated conductors, or optical fibres, or a combination of both, within an enveloping jacket. Either a stranded conductor (single-conductor cable) or a combination of conductors insulated from one another (multiple conductor cable). Small sizes are commonly referred to as stranded wire or as cords.

Bass Boost : Circuitry that boosts the bass response of the amplifier, improving audio reproduction, especially when using inexpensive headphones.

Bastard : Threads, parts, tools, and sizes that are not standard, such as 'bastard nuts,' "bastard plus," "bastard fittings,' and so forth. The term also refers to a standard coarse cut file.

Battery : A device for converting chemical energy into electrical energy. A group of two or more connected voltaic cells.

Battery Capacity : The amount of energy available from a battery. Battery capacity is expressed in ampere-hours.

Battery Backup : A feature of microprocessor supervisory circuits and some power supplies to switch between a main power source and a battery.

Battery Freshness Seal : A feature in microprocessor supervisory circuits which disconnects a backup battery from any downstream circuitry until VCC is applied the first time. This keeps a backup battery from discharging until the first time a board is plugged in and used, and thus preserves the battery life.

Battery Fuel Gauge : A feature or device that measures the accumulated energy added to and removed from a battery, allowing accurate estimates of battery charge level.

Battery Monitor : A feature that monitors the voltage on a battery and indicates when the battery is low. It is usually implemented using a comparator to compare the battery voltage to a specified level. May also include functions such as charging, remaining capacity estimation, safety monitoring, unique ID, temperature measurement, and nonvolatile (NV) parametric storage.

Battery Switchover : A circuit that switches between the higher of a main supply and a backup battery.

Baud : A measurement of speed based on the number of code elements or units per second.

BCD (Binary-coded Decimal) : Representation of a number in which each decimal digit (0-9) is encoded in binary, with four bits per decimal digit.

Beam : A column of light. The main lobe of an antenna radiation pattern.

Beam Splitter : A device for dividing an optical beam into two or more separate beams.

Beam-Lead Chip : Semiconductor chip with electrodes (leads) extended beyond the wafer.

Beam-Power Tube : An electron tube in which the grids are aligned with the control grid. Special beam-forming plates are used to concentrate the electron stream into a beam. Because of this action, the beam-power tube has high power-handling capabilities.

Bearing : An angular measurement of the direction of an object from a reference direction, such as true north.

Bearing Resolution : Ability of a radar to distinguish between targets that are close together in bearing.

Beat Frequency : The frequency produced when two signals are mixed or combined. The beat frequency equals the difference or offset between the two frequencies. The difference between the oscillator frequency and the unknown audio frequency.

Beat-Frequency Oscillator : An additional oscillator used in a receiver when it is receiving a cw signal. It provides an audible tone.

Bed : A very hard wood almost unaffected when exposed to water. Therefore, it is used for keels of ships and flood gates. On account of its hardness it was used for wheels and railway sleepers.

Beech : A very tough and elastic timber used for oars, blocks and pulleys, spokes of wheels and milk-pails. However, it is not very flexible and is difficult to work. Ash bark is used tanning and the ashes of ash is used for potash.

Bell Mouth : The flaring or tapering of a machined hole, usually made at the entrance end because of misalignment or spring of the cutting tool.

Bend Radius : The minimum radius to which a cable, without tensile load, can be bent for its lifetime without causing broken wires or fibers, a localized weakening of the fibers, or a permanent increase in attenuation.

BER (Bit Error Rate) : A measure of the number of erroneous bits which can be expected in a specified number of bits in a serial stream.

BERT (Bit Error Rate (BER) Tester) : A piece of test equipment which determines the bit error rate for a device under test (DUT).

BETA : The ratio of a change in collector current to a corresponding change in base current when the collector voltage is constant in a common-emitter circuit.

BIAS : Difference of potential applied to a vacuum tube or transistor to establish a reference operating level.

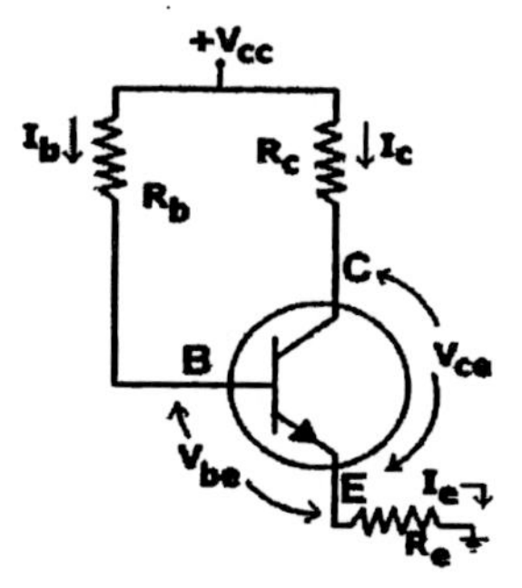

Transistor Bias Circuit

BIAS Current : Current that flows through the base-emitter junction of a transistor and is adjusted to set the operating point of the transistor.

BIAS Resistor : A resistor used to bias a circuit.

Bidirectional : The device accommodates signals traveling either direction though a single channel.

Bifilar Winding : Two or more windings with the wire of each winding alongside the other, matching turn for turn; may be either inductive or noninductive.

Bifunctional Monomer : A monomer unit that has two active bonding positions.

Big-endian : A format for storage or transmission of binary data in which the most significant byte appears first. In a multi-byte value, the byte containing the most significant bit is stored in the lowest memory address and transmitted first, and the byte containing the least significant bit is stored in the highest memory address and transmitted last.

Billet : Rectangular or square bar of raw material.

Binary : A number system that uses a base, or radix, of 2.

Binary Code : A method of representing numbers by a series of digits each of which has only two possible states - either on or off (on = 1, off = 0). Each additional digit doubles the choices available thus four digits (2×2×2×2 = 16 choices) can be used to represent any number from 0 to 15. Digits are usually referred to as "bits" and eight bits equals a "byte" which can represent any number from 0 to 255 (2×2×2×2×2×2×2×2 = 256).

Binary Coded Decimal (BCD) : A four bit binary code that only counts from 0 to 9, because decimal only counts from 0 to 9.

Biochemical Oxygen Demand (BOD) : The amount of oxygen required to oxidise any organic matter present in a water during a specified period of time, usually 5 days.

Biofilm : A film of microorganisms attached to a surface, such as that on a trickling filter, rotating biological contactor, or rocks in natural streams.

Biogeochemical Cycle : The cycle of elements through the biotic and abiotic environment.

Bipolar Junction Transistor : A Bipolar Junction Transistor, or BJT, is a solid-state device in which the current flow between two terminals (the collector and the emitter) is controlled by the amount of current that flows through a third terminal (the base).

Bipolar Signal : A signal that is symmetrical around a reference point, normally ground. A signal that transitions between a positive and negative voltage [logic 1] and a reference point [logic 0]. While a Polar signal might do the same but use the positive voltage as a logic 1 [high] and a negative voltage as a logic 0 [low].

Birch : Strong flexible wood that has many uses. Amongst the uses are furniture making, building as well as being used for agricultural implements in the past.

Bistable : A device that is capable of assuming either one of two stable states. A Flip Flop has two stable output states.

Bistable Multivibrator : A multivibrator that has two stable states. It remains in one of the states until a trigger is applied. It then flips to the other stable state and remains there until another trigger is applied.

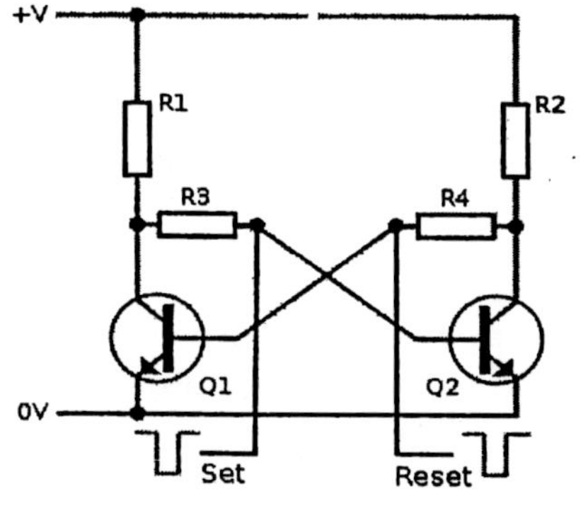

Bistable Multivibrator

BIT : A binary digit.

Bit Banging : A technique which uses the general-purpose ports of a microcontroller to emulate a serial interface standard (I2C, SPI, etc).

BIT Error Ratio : The number of erroneous bits divided by the total number of bits transmitted, received, or processed over some stipulated period.

BIT Rate : In a bit stream, the number of bits occurring per unit time, usually expressed in bits per second.

BIT Slip : In digital transmission, the loss of a bit or bits, caused by variations in the respective clock rates of the transmitting and receiving devices.

Black : The reference color of equipment that passes unclassified information. It normally refers to patch panels or sections of a circuit that work only on unclassified information.

Blade Server : A blade server is a computer system on a motherboard, which includes processor(s), memory, a network connection, and sometimes storage. The blade idea is intended to address the needs of large-scale computing centers to reduce space requirements for application servers and lower costs.

Bleeder Current : The current through a bleeder resistor. In a voltage divider, bleeder current is usually determined by the 10 percent rule of thumb.

Bleeder Resistor : A resistor used to draw a fixed current.

Blind Hole : A hole made in a workpiece that does not pass through it.

Blink Control : Controls the display segment blink rate.

Block : A block of data moved over a bus as a unit in one transfer, using one address cycle. A group of data transferred in a group to reduce bus over-head.

Block Copolymer : A linear copolymer in which identical mer units are clustered in blocks along the molecular chain.

Block Diagram : A diagram in which the major components of an equipment or a system are represented by squares, rectangles, or other geometric figures, and the normal order of progression of a signal or current flow is represented by lines

Block Transfer : The process, initiated by a single action, of transferring one or more blocks of data. Many electrical interfaces use Block Transfers, the VME interface being just one.

Blocked-Grid Keying : A method of keying in which the bias is varied to turn plate current on and off.

Blocking : A condition in an amplifier, caused by over driving one or more stages, in which the amplifier is insensitive to small signals immediately after reception of a large signal. A term describing the state of a semiconductor device or junction which eventually prevents the flow of current.

Blowhole : A defect in a casting caused by trapped steam or gas.

Bluetooth : A technology that allows voice and data connections between a wide range of mobile and stationary devices through short-range digital two-way radio. For instance, it specifies how mobile phones, Wireless Information Devices (WIDs), computers and PDAs interconnect with each other, with computers, and with office or home phones.

BNC Termination : A resistive termination used with BNC connectors, normally 50 ohms.

Termination

Bob Card : A Break-Out-Board is used to 'break out' or separate one style of interface into another via a Printed Wiring Board. For example a BOB card may hold some type of connector [D-sub], but break those lines out to different headers or separate screw terminals. So a normal system cable will still connect to the BOB card using the connector, but each of the lines are now accessible for debug off the individual terminals.

Bode Plot : A graph of a linear, time-invariant system versus frequency.

Body-centered Cubic (BCC) : A common crystal structure that contains atoms located at the corners of a cubic cell and one atom at the cell center position.

Bolometer : A loading device that undergoes changes in resistance as changes in dissipated power occur.

Bond : The permanent joining of metallic parts or circuits assuring electrical continuity and capacity to safely conduct any current likely to be imposed.

Bonding Energy : The energy required to separate two atoms that are chemically bonded to each other.

Bonding Wires : Fine wires connecting the bonding pads of the chip to the external leads of the package.

Boolean : Pertaining to the operations of formal logic.

Boolean Algebra : A system of logic dealing with on-off circuit elements associated by such operators as the AND, OR, NAND, NOR, and NOT functions.

Boost Converter : A power supply that steps an input voltage up (boosts it) to a higher, regulated voltage.

Bootstrap : Often refers to using the output of a step-up converter to drive the main power FET switch, providing more gate drive than the input can supply alone.

Bore : To enlarge and finish the surface of a cylindrical hole by the action of a rotating boring bar (cutting tool) or by the action of a stationary tool pressed (fed) against the surface as the part is rotated.

Boring Bar : A combination tool holder and shank.

Boring Tool : A boring tool is used to enlarge a cylindrical hole, sometimes the tool bit the boring bar and the tool holder are incorporated in one solid piece.

Bounce Time : The time interval between first make of the contact until the uncontrolled making and breaking of the contact ceases.

Boundary Conditions : The two conditions that the E-field and H-field within a waveguide must meet before energy will travel down the waveguide. The E-field must be perpendicular to the walls and the H-field must be in closed loops, parallel to the walls, and perpendicular to the E-field.

Box Wood : Very compact hard wood suitable for machining small parts from. Brass is an alloy of copper and zinc, but often other elements such as aluminium, iron, manganese, tin and lead are added.

Bragg's Law : A relationship that stipulates the condition for diffraction by a set of crystallographic planes.

Braid : A fibrous or metallic group of filaments interwoven to form a protective covering over one or more wires.

Branch : An individual current path in a parallel circuit.

Branched Polymer : A polymer having a molecular structure of secondary chains that extend from the primary chains.

Brazing : A metal joining technique that uses a molten filler metal alloy having a melting temperature greater than about 425°C.

Break : An interruption in a circuit is known as a break. In a switch, the number of breaks refers to the number of points at which the switch opens the circuit; for example, single break and double break.

Break Out Box : A matrix of test points which can be connected in place of an ECU to ease checking of continuity and resistance of a circuit and its components.

Break-before-Make : Contacts which interrupts one circuit before establishing the previous one.

Break-Before-Make : A switch that is configured to break (open) the first set of contacts before engaging (closing) the new contacts. This prevents the momentary connection of the old and new signal paths. Applies to mechanical systems and to solid-state analog multiplexers and switches.

Breakdown **:** The phenomenon occurring in a reverse-biased semiconductor diode. The start of the phenomenon is a transition from a high dynamic resistance to one of substantially lower dynamic resistance.

Breakdown Voltage **:** The breakdown voltage is the maximum instantaneous voltage, including repetitive and non-repetitive transients, which can be applied across a junction in the reverse direction without an external means (circuit) of limiting the current. It is also the instantaneous value of reverse voltage at which a transition commences from a region of high small-signal impedance to a region of substantially lower small-signal impedance.

Bridge Battery **:** A battery intended to provide power to system memory while the main battery is replaced.

Bridge-Tied Load **:** Used in audio applications, the load (a speaker in this case) is connected between two audio amplifier outputs (it "bridges" the two output terminals). This can double the voltage swing at the speaker, compared to a speaker that is connected to ground. The ground-tied speaker can have a swing from zero to the amplifier's supply voltage.

Brightness **:** Although the terms "brightness" and "luminance" are often used interchangeably, they are different. Luminance is the light intensity; brightness is how it is perceived by the human eye.

Brightness Control **:** The name given to the potentiometer used to vary the brightness of a monitor.

Brittle Fracture **:** Fracture that occur by rapid crack propagation and without appreciable macroscopic deformation.

Broach **:** A long, tapered cutting tool with serration's which, when forced through a hole or across a surface, cuts a desired shape or size.

Broadband **:** A transmission medium with enough bandwidth to carry multiple voice, video, or data channels simultaneously. This technique is used, for example, to provide fifty CATV channels on one coaxial cable; or to provide Internet access over cable TV; or to add DSL to a voice-grade telephone line. A common technique is frequency-division. Each channel is modulated to a different frequency band and combined in the transmission medium. It is demodulated to its original frequency at the receiving end. Channels are separated by guardbands (empty spaces) to ensure that each channel will not interfere with its neighbouring channels.

Broadband Lambda Sensor : A type of oxygen sensor that can produce a variable signal output according to a wide range of air/fuel ratios.

Broadside Array : An array in which the direction of maximum radiation is perpendicular to the plane containing the elements.

Bronze : Bronze is an alloy of copper and tin. However, the name is now applied to other alloys that do not contain tin.

Brownout : A condition where the voltage supplied to the system falls below the specified operating range, but above 0V.

Brushes : Sliding contacts, usually carbon, that make electrical connection to the rotating part of a motor or generator.

BSC (Basic Spacing between Centers) : BSC is a term that appears on IC package drawings in reference to dimensions between pins. "Basic" spacing is nominal and can change with conditions. For example, the distance between the rows of pins on a DIP (dual inline package) is BSC because it changes when the auto insertion machine grabs the part, and again when the part is inserted. The BSC dimension, in this case, is the dimension of the hole spacing that the part will fit into, rather than the dimensions of the part itself.

Buck : A "buck" or "step-down" switch-mode voltage regulator is one in which the output voltage is lower than its input voltage. Buck means to resist or reduce (as in "buck the trend"), and hence was used to denote a step-down. Conveniently, it alliterates with the opposite, a boost regulator.

Buck Regulator : A buck converter is a step-down DC to DC converter.

Buck-Boost : A switch-mode voltage regulator in which output voltage can be above or below the input voltage.

Buff : To polish to a smooth finish of high luster with a cloth or fabric wheel to which a compound has been added.

Buffer : A voltage amplifier used between the oscillator and power amplifier. An allocation of memory space for temporary storage. A component or circuit that compensates for a difference in data transfer rates and data processing rates.

Buffer Amplifier : An amplifier that isolates one circuit from another. A Buffer decreases the loading effect on an oscillator by reducing the interaction between the load and the oscillator.

Bull Gear : The large crank gear of a shaper.

Buncher Cavity : The input resonant cavity in a conventional klystron oscillator.

Buncher Grid : In a velocity-modulated tube, the grid that concentrates the electrons in the electron beam into bunches.

Burgers Vector : A vector that denotes the magnitude and direction of lattice distortion associated with a dislocation.

Buried Via : A via hole which does not extend to the surface. A via that only runs between two internal layers.

Burn-In : The process of screening out marginal component parts by exposing them to elevated temperatures and voltage stress. This test is performed to eliminate marginal devices or those with defects resulting from manufacturing aberrations that are evidenced as time and stress dependent failures. Without the burn-in, these defective devices would be expected to result in early lifetime failures under normal use conditions. It is the intent of this test to operate the semiconductor device at specified conditions to reveal electrical failure modes that are time and stress dependent.

Burnishing : The process of finishing a metal surface by contact with another harder metal to improve it. To make smooth or glossy by or as if by rubbing; polish.

Burr : The sharp edge left on metal after cutting or punching. Or. A rotary cutting tool designed to be attached to a drill.

Burst Dimming : Burst Dimming is a method of controlling the brightness of cold cathode fluorescent lamps (CCFL) by turning the lamps on and off at a rate faster than the human eye can detect. The on/off rate is nominally 100Hz to 300Hz. The higher the ratio of on-time to off-time, the brighter the lamps will be. Because of CCFL response times, on-time to off-time ratios of less than 1% are not practical.

Burst Mode : (1) A temporary highspeed data-transfer mode that can transfer data at significantly higher rates than would normally be achieved with nonburst technology. (2) The maximum short-term throughput which a device is capable of transferring data.

Bus : One or more conductors or optical fibers that serve as a common connection for a group of related devices. More information on Bus Topologies. A line shared by more than one device or component so that information can be passed between devices.

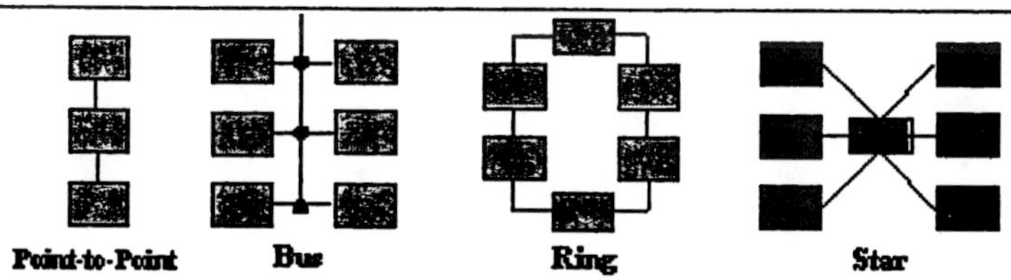

Common Bus Topologies

Bus Bar : A heavy copper strap or bar used to connect several circuits together when a large current carrying capacity is required.

Bushing : A sleeve or a lining for a bearing or a drill jig to guard against wear.

Butterworth Filter : The type of filter with a particula output response of either an active or passive filter.

Bypass Capacitor **:** A capacitor used to transfer [Bypass] unwanted signals out of a circuit; for example, coupling an unwanted signal to ground. A capacitor connected across the power and ground pins of an IC.

Byte **:** A sequence of adjacent bits (usually 8) considered as a unit [character]. Ten bit bytes are also common, as in 8B/10B.

By-wire **:** A "by-wire" term denotes a control system that replaces traditional mechanical or hydraulic linkages with electronic connections between control units that drive electro-mechanical actuators.

C

Cable : An assembly of one or more insulated conductors, or optical fibers, or a combination of both, within an enveloping jacket. Either a stranded conductor (single-conductor cable) or a combination of conductors insulated from one another (multiple conductor cable). Small sizes are commonly referred to as stranded wire or as cords.

Cable Assembly : A cable that is ready for installation in specific applications and usually terminated with connectors.

Cable Carrier : A device that holds a cable.

Cable Clamp : A mechanical clamp attached to the wire entrance of a connector to support the cable or wire bundle, provide stress relief, and absorb vibration and shock.

Cable Harness : A group of wires or ribbons of wiring used to interconnect electronic systems and subsystems.

Cable Sheath : A conductive protective covering applied to cables.

Cable, Coaxial : A cable in which an insulated conductor is centered inside another. The outer conductor is usually a metal braid or metal sheath. Braided cables usually have an outer insulating jacket over the braid. Coaxial cables are used primarily for transmission of RF signals.

Cache : A small portion of memory used as temporary storage, normally having a faster access time than the main memory bank.

Calcination : A high-temperature reaction whereby one solid material dissociates to form a gas and another solid.

Calcium : A silvery white metal which is obtained by the electrolysis of molten calcium chloride. One of the most abundant metal elements on Earth, but it never occurs as the pure element.

Caliper : A measuring instrument used to measure the distance between two points or the inside or outside dimensions of an object.

Cam : A device for converting regular rotary motion to irregular rotary or reciprocating motion. Sometimes the effect of off-center lathe operations or the cam in an engine used to operate valves.

Can Bus : A differential 2-wire interface running over either a Shielded Twisted Pair (STP), Un-shielded Twisted Pair (UTP), or Ribbon cable. A number of different Protocols are used with CAN bus. The CanBus is primarily used as an Automotive Bus, but is also one of the major Field Buses.

Capacitance Unbalance : A measure of the inequality of capacitance between the conductors of two adjacent pairs in a multi-pair cable.

Capacitance : The charge-storage ability of a capacitor, defined as the magnitude of charge stored on either plate divided by the applied voltage.

Capacitive Crosstalk : A phenomenon where a signal on one line/trace is capacitively coupled to an adjacent line/trace.

Capacitive Discharge Ignition (CDI) : A type of ignition system in which a capacitor is charged to a high voltage and discharged through the coil primary to produce the required HT voltage. Spark duration is usually very brief.

Capacitive Reactance : The opposition, expressed in ohms, offered to the flow of an alternating current by capacitance. The symbol for capacitive reactance is XC.

Capacitor Filter : This filter is used on extremely high-voltage, low-current power supplies and also where the ripple frequency is not critical.

Capacitor **:** An electrical device capable of storing electrical energy in an electrostatic field.

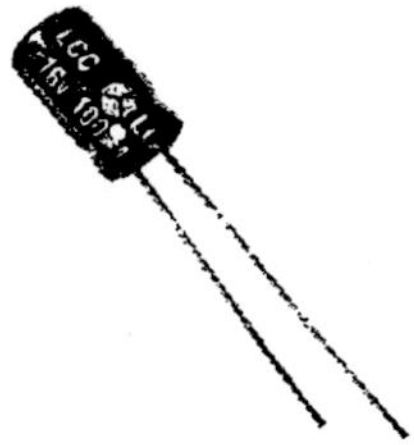

Capacitor

Capacitor-Start Motor **:** A type of single-phase, ac induction motor in which a starting winding and a capacitor are placed in series to start the motor. The values of XC and R are such that the main-winding and starting-winding currents are nearly 90 degrees apart and the starting torque is produced as in a two-phase motor.

Carbide Tools **:** Lathe cutting tools to which carbide tip inserts have been brazed or screwed, to provide cutting action on harder materials than the high speed cutters are capable of.

Carbon Dioxide **:** A colc rless gas exhaled by animals and absorbed by plants. It is formed by the complete combustion of carbon.

Carbon Microphone **:** A microphone in which sound waves vary the resistance of a pile of carbon granules. May be single-button or double-button.

Carbon Monoxide (CO) Analysis. Measurement of CO concentration in exhaust to determine mixture strength when richer than stoichiometric. Expressed as percentage of total gas volume. A typical full throttle mixture would produce 3-5 % CO.

Carbon Steel **:** A broad term applied to tool steel other than highspeed or alloy steel.

Carbonaceous Biochemical Oxygen Demand (CBOD) **:** The amount of oxygen required to oxidise any carbon containing matter present in a water.

Carbonising : The process of adding carbon to the outer surface of steel to improve its quality by heat treating it in contact with a carbonaceous material.

Carbonyl : A functional group with an oxygen atom double bonded to a carbon atom.

Carbonated : Used extensively as a grinding compound and in abrasive wheels.

Carburising : The process by which the surface carbon concentration of a ferrous alloy is increased by diffusion from the surrounding environment.

Cardbus : One of the different types of PCMCIA interfaces. Implements the 32-bit PCI bus standard into the PCMCIA form factor.

Carriage : A principal part of a lathe that carries the cutting tool and consists of the saddle, compound rest. And apron.

Carrier Frequency : The base frequency of a transmitted electromagnetic pulse or wave on which information can be imposed by varying the signal strength, varying the base frequency, varying the wave phase, or other means. The frequency of an unmodulated transmitter output.

CAS (Column-Address-Strobe) : The signal that tells the DRAM to accept the given address as a column-address; used with RAS and a row-address to select a bit within the DRAM.

Cascamite : Trade name for a urea-formaldehyde glue. This is a waterproof resin based glue that comes as a white powder and is mixed two parts water to one part powder in a glass container. Note that it must be stirred thoroughly until it has a consistency of cream.

Castigated Nut : A nut with grooves cut entirely across the top face used with a split pin to lock the nut and stop it from undoing. Often used in safety critical applications.

Cathodic Protection : A means of corrosion prevention whereby electrons are supplied to the structure to be protected from an external source such as anoother more reactive metal or a dc power supply.

Cavity Wavemeter : An instrument used to measure microwave frequencies.

CCD (Charge Coupled Device) : One of the two main types of image sensors used in digital cameras. When a picture is taken, the CCD is struck by light coming through the camera's lens. Each of the thousands or millions of tiny pixels that make up the CCD convert this light into electrons. The accumulated charge at each pixel is measured, then converted to a digital value. This last step occurs outside the CCD, in an analog-to-digital converter (ADC).

CCFL (Cold Cathode Fluorescent Lighting) : Often used as a backlight for LCD displays.

CCFT (Cold Cathode Fluorescent Tube) : Often used as a backlight for LCD displays.

CDMA (Code Division Multiple Access) : A digital cellular technology that uses spread-spectrum techniques. Unlike GSM and other competing systems that use TDMA, CDMA does not assign a specific frequency to each user. Instead, every channel uses the full available spectrum. Individual conversations are encoded with a pseudo-random digital sequence.

Cedar : A durable wood used in ancient buildings. Easy to work and strongly scented.

Cell : A unit of varying dimensions in a landfill which is isolated from the environment by 6 to 12 inches of soil cover. A cell is one day's waste or less. A cell is covered with soil at the end of each day.

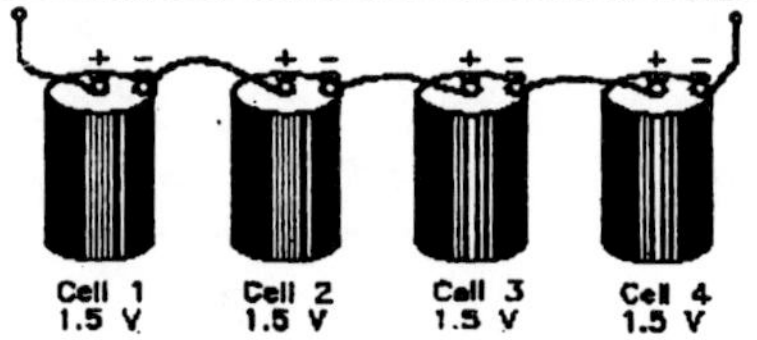

Battery Cell

Cement : A substance that can be used to build together aggregates of sand or stone into a cohesive structure. May be a single compound or a mixture. May be hydraulic set, air set or chemical set.

Centres : The centre of rotation of an object. A centre as used in a head stock or tail stock.

Centrifugal Starting Switch : A mechanism that disconnects the starting circuit of a motor when the rotor reaches approximately 75% of operating speed.

Ceramic : Inorganic, non-metalllic products for which the interatomic bonding is predominantly ionic.

Cermet : A composite material consisting of a combination of ceramic and metallic materials.

Ch. to Ch. Skew (Ps Max) : Channel-to-channel skew. A signal on one channel has a different phase than the same signal on another channel (delayed/skewed). This is measured in picoseconds, max.

Changeover Valve : Solenoid operated valves used in combination to control fuel feed and return selection on twin tank systems.

Channel : A carrier frequency assignment, usually with a fixed bandwidth.

Channel Associated Signaling : Some communications protocols include "signaling" functions along with data. Channel Associated Signaling protocols include signaling in the data channel (as opposed to a dedicated signaling channel). Also called Robbed Bit Signaling.

Character : A letter, digit, or other symbol that is used as part of the organization, control, or representation of information.

Characteristic Impedance : The ratio of voltage to current at any given point on a transmission line represented by a value of impedance. A measure of the electromagnetic coupling between the signal conductor and its return path. The impedance of a circuit that, when connected to the output terminals of a uniform transmission line of arbitrary length, causes the line to appear infinitely long. A uniform line terminated in its characteristic impedance will have no standing waves, no reflections from the end, and a constant ratio of voltage to current at a given frequency at every point on the line.

Charge : Represents electrical energy. A material having an excess of electrons is said to have a negative charge. A material having a shortage of electrons is said to have a positive charge.

Charge Cycle : The period of time that a capacitor in an electrical circuit is storing a charge.

Charge Injection : A parameter pertinent to analog switches. As an analog switch turns on and off, a small amount of charge can be capacitively coupled (injected) from the digital control line to the analog signal path.

Charge Pump : A power supply which uses capacitors to store and transfer energy to the output, often stepping the voltage up or down. Charge is transferred from one capacitor to another under control of regulator and switching circuitry.

Charge Termination Method : Method the battery charger uses to determine when to terminate the charging cycle.

Chassis : An equipment case. An equipment enclosure used to hold circuit boards or other components.

Chassis Ground : The potential of an equipment chassis. The attached graphic shows how to connect a printed wire board [PWB] ground to chassis ground.

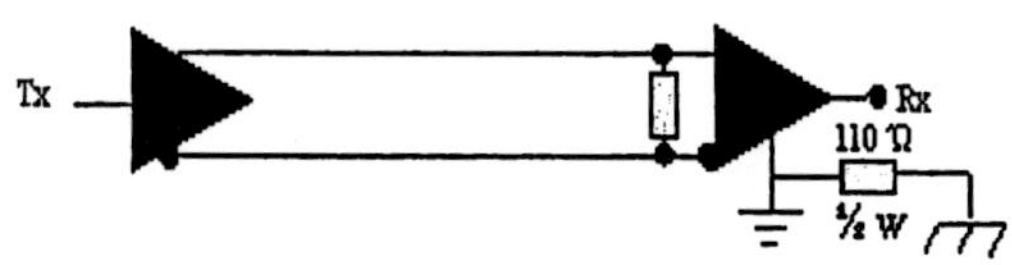

Chassis Ground

Chatter, Contact : The undesired opening of mating contacts resulting from uncompensated AC operation, or external shock and vibration.

Check BIT : A bit, such as a parity bit, derived from and appended to a bit string for later use in error detection and possibly error correction.

Check-Sum : A number that has been calculated as a function of some message, by adding up the bytes in the message. The [message] data is summed with out regard overflow. Perhaps this was what early checksums were.

Chemical Fixation (or Stabilisation/solidification) : A term for several different methods of chemically immobilising hazardous materials into a cement, plastic, or other matrix.

Chemical Oxygen Demand (COD) : The amount of oxygen required to oxidise any organic matter in the water using harsh chemical conditions.

Chemical Set Cement : A cement that sets through reaction or precipitation. Often subjected to a high temperature during manufacture or use.

Chestnut : Very durable and much like oak. Comparitavely easy to work and does not shrink much when drying

Chip : (1) Integrated circuit : A semiconductor device that combines multiple transistors and other components and interconnects on a single piece of semiconductor material. (2) Encoding element, in Direct-Sequence Spread Spectrum systems.

Chip Carrier : A low-profile four-sided (rectangular) part package, whose semiconductor chip cavity or mounting area is a large fraction of the package size.

Chip-Enable Gating : A feature in microprocessor supervisory circuits which prevents the writing of erroneous data when power falls outside of spec. When the main power-supply voltage is below the minimum safe-operating limit, the feature disconnects the chip-enable signal path from the host microprocessor or microcontroller.

Chlorofluorocarbons **:** Synthetic organic compounds used for refrigerants, aerosol propellants (prohibited in the U.S.), and blowing agents in plastic foams.CFCs migrate to the upper atmosphere destroying ozone and increasing global warming. Typical atmospheric residence times are 50 to 200 years.

Choke **:** An inductor used to impede the flow of pulsating dc or ac by means of self-inductance. A choke is designed to have a high reactance to a particular frequency. Common-mode chokes are used to protect from EMI. High frequency chokes may be called ferrite chokes or choke baluns.

Choke

Choke Joint **:** A joint between two sections of waveguide that provides a good electrical connection without power losses or reflections.

Chrominance **:** The color portion portion of a composite video signal. Forms a complete picture once combined with the luminance component.

Circuit Breaker **:** A device designed to open and close a circuit by nonautomatic means and to open the circuit automatically on a predetermined over-current without injury to itself when properly applied within its rating.

CISC (Complex Instruction Set Computer) **:** Computer hardware designed to support complex instructions, as opposed to RISC (reduced instruction set computer) architecture.

Clarifier **:** A tank in which quiescent settling occurs, allowing solid particles suspended in the water to agglomerate and settle to the bottom of the tank. The solids resulting from the settling being removed as a sludge.

Class A Amplifier : The simplest type of amplifier, class A amplifiers are those in which the output transistors conduct (i.e. do not fully turn off) irrespective of the output signal waveform. This type of amplifier is typically associated with high linearity but low efficiency.

Class AB Amplifier : Class AB amplifiers combine Class A and Class B to achieve an amplifier with more efficiency than Class A but with lower distortion than class B. This is achieved by biasing both transistors so they conduct when the signal is close to zero (the point where class B amplifiers introduce non-linearities). The transistors transition to class B operation for large excursions. So, for small signals both transistors are active, acting like a class A amplifier. For large signal excursions, only one transistor is active for each half of the waveform, acting like a class B amplifier.

Class B Amplifier : Class B amplifiers are those in which the output transistors only conduct during half (180 degrees) of the signal waveform. To amplify the entire signal two transistors are used, one conducting for positive output signals and the other conducting for negative outputs. Class B amplifiers are much more efficient than class A amplifiers, but have high distortion due to the crossover point when the two transistors transition from on to off.

Class C Amplifier : A class C amplifier is a form of switching amplifier in which the transistors are on for less than a half cycle (less than 180 degrees) often, much less. For instance, the transistor may be on only during the top 10% of the signal excursion, delivering just a pulse. Class C amps are very efficient because the transistors are off most of the time and when they are on, they are in full conduction. They deliver high distortion and are often used in RF circuits, where tuning circuits restore some of the original signal and reduce distortion. They are also used in low-fidelity applications where the distortion is not important, such as a siren speaker driver.

Class D Amplifier : Class D amplifiers are those that output a switching waveform, at a frequency far higher than the highest audio signal that needs to be reproduced. The low-pass filtered, average value of this waveform corresponds to the actual required audio waveform. Class D amplifiers are highly efficient (often up to 90% or higher) since the output transistors are either fully turned on or off during operation. This completely eliminates the use of the linear region of the transistor which is responsible for the inefficiency of other amplifier types. Modern Class D amplifiers achieve fidelity comparable to class AB.

Class G Amplifier : Class G amplifiers are similar to class AB amplifiers except they use two or more supply voltages. When operating at low signal levels, the amplifier uses a lower supply voltage. As the signal level increases, the amplifier automatically picks the appropriate supply voltage. Class G amplifiers are more efficient than class AB amplifiers since they use the maximum supply voltage only when required, while a class AB amplifier always uses the maximum supply voltage.

Class H Amplifier : Class H amplifiers modulate the supply voltage to the amplifier output devices so that it is never higher than necessary to support the signal swing. This reduces dissipation across the output devices connected to that supply and allows the amplifier to operate with an optimised class AB efficiency regardless of output power level. Class H amplifiers are generally more complex than other designs, with extra control circuitry required to predict and control the supply voltage.

Click/Pop Reduction : A feature that eliminates "clicks" and "pops" unwanted transient noise signals during power-up, shutdown, connection, etc.

Click-and-Pop : Click-and-pop refers to the unwanted transient signals in the audio band that are reproduced by the head phone and/or speaker when the audio device driving it is either: powered up (power applied) : powered down (power removed) brought out of shutdown (power applied previously) forced into shutdown (power still applied).

Climatology : The study of the climate, how the earth's atmosphere performs over long periods of time.

Clock and Data Recovery : The process of extracting and reconstructing clock and data information from a single-wire/channel, serial data stream.

Clock Jitter : A periodic waveform (especially a clock) is expected to cross certain thresholds at precisely timed moments. Variations from this ideal are called jitter.

Clock Throttling : Reducing the frequency or duty-cycling the clock of an integrated circuit usually for the purpose of reducing heat generation.

Closed Loop Operation : A condition in which a signal derived from an output is looped back to correct the input. Usually refers to operation where the signal from the Lambda sensor is used to make fuelling corrections. Also known as Lambda Feedback.

Closure : The act of preparing a landfill for long term inactivity, including placement of a cover over the landfill to prevent infiltration of surface water.

CMOS : Complementary metal-oxide semiconductor technology in which p- and N.-channel MOS transistors are used in tandem.

CMRR (Common Mode Rejection Ratio) : The ability of a differential amplifier to not pass (reject) the portion of the signal common to both the + and - inputs.

Coagulation : Particle destabilisation to enhance agglomeration.

Coaxial Line : A transmission line in which one conductor is located at the centre of a metal tube which acts as the second conductor.

CODEC : Short for compressor/decompressor, a codec is any technology for compressing and decompressing data. Codecs can be implemented in software, hardware, or a combination of both.

Coherent Sampling : Describes the sampling of a periodic signal, where an integer number of its cycles fits into a predefined sampling window.

Coil : An inductive device made by looping turns of wire around a core.

Through-Hole Coils

Cold Start Injection : A simple system operating independently of the ECU and main injectors, which only functions briefly at low temperatures, spraying a fine mist of fuel into the inlet tract whilst the starter is engaged. Rarely used since the improved spray patterns of later injectors made possible the simpler method of firing the main injectors at an increased rate during cranking.

Cold Start Injector. A special injector sometimes used to provide a fine fuel mist in the plenum chamber to assist fire up while the starter is cranking the engine over.

Cold Working : The plastic deformation of a metal at a temperature below that at which it recrystallises.

Colloids : Small particles which have a negligible settling velocity. These particles have a very small mass so gravitational force is low compared to surface frictional forces. Typical colloidal sizes range from 10-3 mm to 1 mm.

Color Subcarrier : A modulated carrier, added to a television signal, to carry the color components. Examples: In NTSC television, a 3.579545MHz color subcarrier is quadrature-modulated by two color-difference signals and added to the luminance signal. The PAL television standard uses a subcarrier frequency of 4.43362MHz.

Colour Temperatures : The colour of a piece of steel may be used as a guide to it's temperature. These are particulalrly useful when hardening or tempering steel.

COMB Filter : An N-Path filter response which contains a sequence of equally spaced passbands forming a comb.

Combination Array : An array system that uses the characteristics of more than one array.

Combination Circuit : A series-parallel circuit.

Combination Peaking : A technique in which a combination of peaking coils in series and parallel (shunt) with the output signal path is used to improve high-frequency response.

Combinatorial Logic : A logic subsystem that contains no memory who's output is a function of the present input.

Commercial-Off-The-Shelf : [COTS] item is defined as any item that can be purchased and does not need to be designed.

Common Base : The Base element is common to both the input and output circuit, normally grounded.

Common Carrier : In the telecommunications area, the term used to describe a telephone company.

Common Collector : A transistor circuit which has the collector of the transistor grounded or common between the input and output of the transistor circuit.

Common Identities Law : In Boolean algebra this law states that anytime the expression A(A + B) = AB or A + AB = A + B appears, it can immediately be simplified to AB without going through the process of using the distributive law, complementary law, or the law of union to simplify.

Common Mode Conversion Ratio : A measure of the balance of the two conductors of a pair with respect to earth.

Common-Base Detector : An amplifying detector in which detection occurs in the emitter-base junction and amplification occurs at the output of the collector junction.

Common-Emitter Detector : Often used in receivers to supply detected and amplified output. The emitter-base junction acts as the detector.

Common-Mode Signals : Common-mode signals are identical signal components on both the + and - inputs of a differential amplifier or instrumentation amplifier. A common example is in a balanced pair, where a noise voltage is induced in both conductors. Another example is where a DC component is added (e.g. due to a difference in ground between the signal source and the receiver). In an ideal differential amp, the common-mode element is cancelled out, since the differential (+ and -) inputs should subtract out the identical components.

Commutation **:** The act of a commutator in converting generator output from an ac voltage to a dc voltage.

Commutative Law **:** In Boolean algebra this law states that changing the order of the terms in an equation will not affect the value of the equation. Example: A + B = B + A; A • B = B • A.

Commutator **:** A mechanical device that reverses armature connections in motors and generators at the proper instant so that current continues to flow in only one direction. In effect, the commutator changes ac to dc.

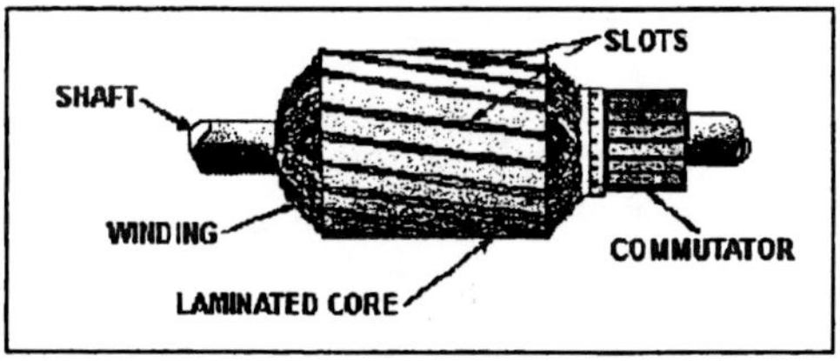

Motor Commutator

CompactPCI **:** Compact Peripheral Component Interface. An industrial version of the PCI standard [Peripheral Component Interface]. CompactPCI uses the Euro board format/size, but uses the same electrical standard as the consumer PCI interface bus.

Compander **:** An acronym for COMpressor-exPANDER. A circuit that compresses and then expands the signal in amplitude.

Compander **:** Signal processing technique which uses both compression and expansion to improve dynamic range and signal-to-noise ratio. A signal is passed through a non-linear transformation prior to transmission. A reverse of this transformation occurs at reception. The transformation is such that quiet portions are boosted and loud portions reduced. Noise is reduced because the quiet signals are louder, compared to the noise in the transmission channel. Used in digital, PCM, transmissions as well as analog applications. Dolby is a common example of a compander-based noise reduction system.

Comparator : A comparator is a device that accepts two analog inputs, compares the inputs, and produces a binary output that is a function of which input is higher. If the non-inverting (+) input is greater than the inverting (-) input, then the output goes high. If the inverting (-) input is greater than the non-inverting (+) input, then the output goes low. When described that way, the comparator resembles a 1-bit ADC. A simple comparator can be achieved using an op amp without negative feedback. Its high voltage gain enables it to resolve very small differences in input voltage. But op amps used this way are generally slower than comparators and lack special features, such as hysteresis and internal references.

Comparator Propagation Delay : This is the lag between the input crossing the comparator threshold, and the output changing states.

Compensating Windings : Windings embedded in slots in pole pieces, connected in series with the armature, whose magnetic field opposes the armature field and cancels armature reaction.

Compensation : The process of overcoming the problems associated with high frequencies in an amplifier.

Complement : A number or state that is the opposite of a specified number or state. The negative of a number is often represented by its complement.

Complement Number : A number that when added to another number gives a sum equal to the base of the number system of operation. For example, in the decimal number system, the complement of 1 is 9.

Complementary (Secondary) Colors of Light : The colors of light produced when two of the primaries are mixed in overlapping beams of light. The complementary colors of light are magenta, yellow, and cyan.

Complementary Law : In Boolean algebra this law states that the logical addition of a quantity and its complement will result in 1 and the logical multiplication of a quantity and its complement will result in a product of 0.

Complementary Metal Oxide Semiconductor : [CMOS] A type of Integrated Circuit structure.

Complex Wave : A waveform other than a sine wave. A wave that is produced by combining two or more pure tones at the same time.

Complexation : The ionic bonding of one or more central ions or molecules by one or more surrounding ions or molecules.

Compliant Pin : A pin used with a Printed Circuit Board [PCB] as an interconnect. The pin is press-fit and solder-free. In most cases the pin[s] are the board mount side of a connector.

Component : A part of a mixture or solution.

CompoNet : CompoNet is a four-wire, industrial, bus with a master-slave architecture. It is used at the lower network levels to transmit bit or word information, such as for use in sensors and actuators. Up to 256 slaves are supported on a bus. Data rates of 93.75kbps to 4Mbps and network lengths up to 1500 meters with repeaters are possible. CIP is used as the underlying protocol.

Composite : A material brought about by combining materials differing in composition or form on a macroscale for the purpose of obtaining specific characteristics and properties. The constituents retain their identity such that they can be physically identified and they exhibit an interface between one another.

Composting : The controlled aerobic degradation of organic wastes into a material which can be used for landscaping, landfill cover, or soil conditioning.

Compound : A substance composed of two or more elements.

Compound Wound Motor : Operation is similar to shunt motor : Powerful starting torque from main circuit winding

Compound-Wound Motors And Generators : Machines that have a series field in addition to a shunt field. Such machines have characteristics of both se. ʹ ɜs- and shunt-wound machines.

Compression Settling : Settling which occurs in the lower reaches of clarifiers where particle concentrations are highest. Particles can settle only by compressing the mass of particles below.

Compression Waves : Longitudinal waves that have been compressed (made more dense) as they move away from the source.

Computer : A data processing unit that can perform computation, including numerous arithmetic or logic operations, without intervention by a human.

Computer

Concrete : A composite material consisting of aggregate particles bound together in a solid body by a cement.

Concurrent : Pertaining to the occurrence of two or more events or activities within the same specified interval of time.

Condensation Polymerisation : The formation of polymers by an intermolecular reaction involving at least two monomer species, usually with the production of a low molecular weight by-product such as water.

Conductance : The ability of a material to conduct or carry an electric current. It is the reciprocal of the resistance of the material and is expressed in mhos or siemens.

Conducted Interference : Interference that occurs because of inductive or capacitive coupling.

Conduction : This is the process by which heat flows from the hotter region of a substance to the colder region without there being any net flow of the material itself.

Conduction Band : The lowest-lying electron energy band that is not completely filled with electrons.

Conduction : The transfer of heat in a substance from a higher temperature to a lower temperature. Heat Conduction is the most effective method of moving heat from the junction to the case and the case to a heat-sink [if installed] in semiconductors.

Conductivity : The ease with which a substance transmits electricity.

Conductor Losses : Power losses due to the resistance of conductors.

Conductor : A material with a large number of free electrons. A material that easily permits electric current to flow. Any wire, cable, or substance capable of carrying an electrical current.

Conformal Coating : An insulative coating that conforms to the configuration of the object being coated.

Congruent Transformation : A transformation of one phase to another that does not involve any change in composition.

Conical Scanning : Scanning in which the movement of the beam describes a cone, the axis of which coincides with that of the reflector.

Constant Energy Ignition : An ignition system employing an amplifier module which can control current flow through the coil to a predetermined level without need for a ballast resistor. The coil energy is therefore constant over a wide range of engine speeds.

Constant Voltage Source : A voltage source shall be considered constant if doubling the generator impedance does not produce a change in the parameter being measured that is greater than the required precision of the measurement.

Contact Bounce : When a mechanical switch or relay closes, the switch elements will often bounce, even if only briefly, before making final contact. This is of consequence if downstream elements are sensitive to the switching transients. A contact debouncing circuit is often used to remove the transients.

Contact Discharge : An ESD test method where the ESD generator makes direct contact with the device under test (DUT).

Convection : The heat is transfered from the solid or liquid by the surrounding gas. If the flow of gas away from the object starts by itself it is called natural convection. However, if the gas is forced past the object it is called forced convection.

Conversion of Units : Conversion factors from Imperial to Metric to Imperial for length, area, volume, mass.

Coolant Sensor : A two terminal device, protruding into the engine coolant system, usually near the thermostat, the resistance of which changes with temperature. The resistance normally falls with rising temperature hence these devices are sometimes correctly described as 'negative temperature coefficient (n.t.c.) Thermostats'.

Coordination Number : The number of atomic or ionic nearest neighbours.

Coplanar Line : A line which is in the same plane as another line. Any two intersecting lines must lie in the same plane, and therefore be coplanar.

Copolymer : A polymer that consists of two or more dissimilar mer units in combination along its molecular chains.

Copper Losses : Power losses due to the resistance of copper conductors.

Copper : Noted for being ductile and having good electrical and thermal conductivity. It forms the basis of a number of alloys.

Corrosion : Deteriorative loss of a metal as a result of dissolution environmental reactions.

Corrosive Waste **:** A waste that is outside the pH range of 2 to 12.5 or a waiste that corrodes steel at a rate greater than 6.35 mm (0.25 in) per year. One of EPA's four hazardous waste properties.

Covalent Bond **:** A primary interatomic bond that is formed by the sharing electrons between neighbouring atoms.

Cranking Enrichment **:** Extra fuel provided by the ECU via the normal injectors whilst the starter is engaged to assist fire up. Not the same as that provided by the cold start injector.

CRC (Cyclic Redundancy Check) **:** A check value calculated from the data, to catch most transmission errors. A decoder calculates the CRC for the received data and compares it to the CRC that the encoder calculated, which is appended to the data. A mismatch indicates that the data was corrupted in transit. Depending on the algorithm and number of CRC bits, come CRCs contain enough redundant information that they can be used to correct the data.

Creep **:** The time-dependent permanent deformation that occurs under stress; for most materials it is important only at elevated temperatures.

Crosslinked Polymer **:** A polymer in which adjacent linear molecular chains are joined at various positions by covalent bonds.

Crossover **:** In an output stage (or similar amplifying stage which uses one device to pull the signal up and another to pull the signal down), the region in which the high-side device is turning on and the low-side device is turning off, or vice versa.

Crosstalk **:** The unwanted coupling of signals between different pairs in a cable.

Crosstalk Distance **:** The decibel difference between the level of a wanted signal on a transmission line and an unwanted signal caused by crosstalk.

Crosstalk Ratio **:** The ratio of crosstalk signal level to the level of the originating source of the crosstalk.

Crowbar Circuit **:** A crowbar circuit is a power supply protection circuit that rapidly short-circuits ("crowbars") the supply line if the voltage and/or current exceeds defined limits. In practice, the resulting short blows a fuse or triggers other protection, effectively shutting down the supply. It is usually achieved by an SCR or other silicon device, or by a mechanical shorting device. Probably named for the concept of using a big metal bar to mechanically provide the short circuit, as might be used done in a high-current application; or from the appearance of a crowbar circuit's I-V curve.

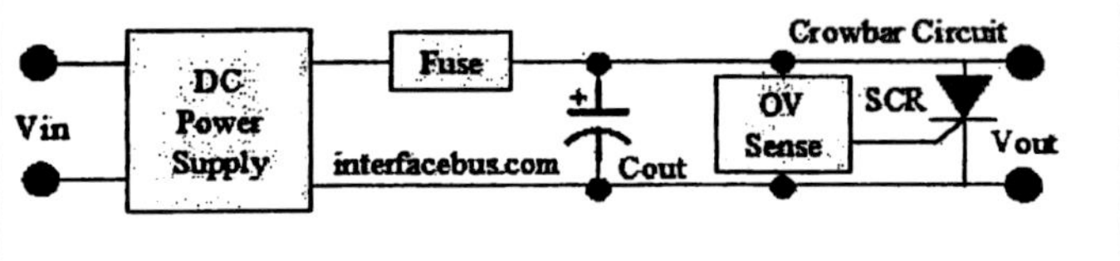

Crowbar Circuit

Cryptanalysis **:** The art and science of breaking encryption or any form of cryptography.

Crystal Structure **:** For crystalline materials, the manner in which atoms or ions are arrayed in space. It is defined in terms of the unit cell geometry and the atom positions within the cell.

Crystal System **:** A scheme by which crystal structures are classified according to unit cell geometry.

Crystalline **:** The state of a solid material characterised by a periodic and repeating three-dimensional arrays of atoms, ions, or molecules.

Crystallinity **:** For polymers, the state wherein a periodic and repeating atomic arrangement is achieved by molecular chain alignment.

Crystallite **:** A region within a crystalline polymer in which all the molecular chains are ordered and aligned.

CSP (Chip Scale Package) **:** An IC packaging technology in which solder balls take the place of pins, making the smallest package available. When heated, the solder balls alloy to matching pads on the circuit board.

Curie Temperature : That temperatue above which a ferromagnetic or ferrimagnetic material becomes paramagnetic.

Current Mode Feedback : An alternative op amp topology usually used in highspeed amplifiers. It is sensitive to feedback impedance, and cannot be used as an integrator.

Current-Mode Controller : A DC-DC switching regulator which regulates its output voltage by varying the peak inductor current on a cycle-by-cycle basis to output a regulated voltage despite variations in load-current and input-voltage.

Current-Sense Amplifier : An amplifier that measures current by measuring the voltage drop across a resistor placed in the current path. The current sense amp outputs either a voltage or a current that is proportional to the current through the measured path.

Currie Point : This is the temperature at which steel/iron loses its magnetic properties. For pure iron this is about 770°C.

Cutting Speeds : Tables of cutting speeds are useful when starting out in engineering or tackling a new material. A few tables are given in the following links for some typical materials.

D

D (deci) : The prefix d or deci indicates the unit should be multiplied by 10^{-1}.

D Flip Flop : A style of flip flop who's output follows the input when the device is clocked. The Truth Table for a D-Type flip flop is shown to the right. The output changes on the edge of the clock. The most common IC is the 7474; Dual Positive-Edge-Triggered D-Type Flip-Flops with Set and Reset. IC Vendors.

Input	Outputs	
@ t_n	@ t_n +1	
D	Q	$\overline{Q}$
L	L	H
H	H	L

Truth Table

D/A Converter : Digital-to-analog converter or DAC, that receives digital data (a stream of numbers) and outputs a voltage or current proportional to the value of the digital data.

D'arsonval Meter Movement : The permanent-magnet moving-coil movement used in most analog meters. Note that most meters use digital displays.

Da (deca) : The prefix d or deca indicates the unit should be multiplied by 10.

Daisy Chain : A method of propagating signals along a bus in which the devices are connected in series and the signal passed from one device to the next. The daisy chain scheme permits assignment of device priorities based on the electrical position of the device on the bus.

Dallastat : Trademark for Dallas Semiconductor's line of digital rheostats (digital potentiometers).

Damped Wave : A sinusoidal wave in which the amplitude steadily decreases with time. Often associated with energy loss.

Damping : The process of smoothing out oscillations. The progressive decay with time in the amplitude of the free oscillations in a circuit. The progressive diminution with time of certain quantities characteristic of a phenomenon. In a meter, this process is used to keep the pointer of the meter from overshooting the correct reading. A mechanical or electrical technique used in synchro receivers to prevent the rotor from oscillating or spinning. Damping is also used in servosystems to minimise overshoot of the load.

Darlington Transistor : Actually a Darlington transistor is a package containing a pair of transistors coupled together in a configuration which multiplies the gain (or amplification) so that a very large current can be controlled by a very small one. Darlingtons are widely used as output transistors for driving coils and injectors.

Data Acquisition System : System which acquires data, generally by digitising analog channels and storing the data in digital form. These systems can be standalone or married to a computer and can acquire multiple channels of data.

Data Converter : An electronic circuit that converts analog signals to digital, or vice-versa. An analog signal is a continuously varying voltage or current. Its digital counterpart is a stream of digital numbers, each representing the amplitude of the analog signal at a moment in time.

Data Encryption Standard (DES) : A cryptographic algorithm for the protection of unclassified computer data and published by the National Institute of Standards and Technology in Federal Information Processing Standard Publication 46-1.

Data Processing : The execution of a systematic sequence of operations performed upon data. Synonymous with information processing.

Data Transfer Rate : The average number of bits, characters, or blocks per unit time passing between corresponding equipment in a data transmission system.

Data Transmission : The transfer of information from one place to another or from one part of a system to another. The sending of data from one place to another by means of signals over a channel.

DB (Decibels) : A method for specifying the ratio of two signals. DB = 10 times the log of the ratio of the power of the two signals. This is equal to 20 times the ratio of their voltages, if the signals are driving equal impedances. Decibels are also used to describe a signal level by comparing it to a reference level. The reference is usually defined as 0dB and the dB value of the signal is 10 times the log of the signal's power over that of the reference. A letter is sometimes added to signify the reference. For instance, dBm is relative to 0 dBm = 1mW.

dBm : A unit that defines a signal level by comparing it to a reference level. The reference level of 0dBm is defined as 1mW. The signal level in dBm is 10 times the log of the signal's power over that of the 0dBm reference.

DBS (Direct Broadcast Satellite) : A system which broadcasts directly from satellite to the subscriber (end user). Prominent examples in the US are Direct TV and Dish network.

DC (Direct Current) : DC is the unidirectional flow of electric charge. Direct current is produced by such sources as batteries, thermocouples, solar cells, and commutator-type electric machines of the dynamo type. Direct current may flow in a conductor such as a wire, but can also be through semiconductors, insulators, or even through a vacuum as in electron or ion beams.

DC-DC : Any of the family of switch-mode voltage regulators, these devices use an inductor to store and transfer energy to the output in discrete packets, resulting in highly efficient power conversion.

DC-DC Controller : A DC-DC converter (switch-mode power supply) in which the power switch (usually a power MOSFET) is external to the IC.

DDR Memory (Double Data Rate Synchronous) : A clock is used to read data from a DRAM. DDR memory reads data on both the rising and falling edge of the clock, achieving a faster data rate. Often used in notebook computers because it also consumes less power.

DDS (Direct Digital Synthesis) : DDS is a method for digitally generating analog waveforms, such as sine waves (modulated or not) or arbitrary waveforms. In the most straightforward realisation, a digitised sample of the waveform is stored and the values are clocked out to a D/A converter. Varying the clock rate changes the frequency. Variations in rate and changes to a gain factor can modulate the signal.

Dead Centre : A centre that does not rotate found on the tail stock of a lathe. *Or*. The exact centre of an object.

Debounce : Electrical contacts in mechanical pushbutton switches often make and break contact several times when the button is first pushed. A debouncing circuit removes the resulting ripple signal, and provides a clean transition at its output.

Decibel : A logarithmic unit used for expressing the ratio of two powers.

Decimal : Pertaining to the number representation system with a radix of ten.

Decomposers : Organisms which utilise energy from wastes or dead organisms. Decomposers complete the cycle by returning nutrients to the soil or water and carbon dioxide to the air or water.

Decoupling Capacitor : A capacitor used to transfer unwanted signals out of a circuit; for example, coupling an unwanted signal to ground. Also called a Bypass Capacitor. A capacitor that De-Couples an IC from its supply voltage by supplying a voltage charge when required reducing the time lag from a distant power supply.

Deemphasis : In FM transmission, the process of restoring (after detection) the amplitude-vs.frequency characteristics of the signal.

Deenergized. Free from any electrical connection to a source of potential difference and from electric charge; not having a potential different from that of the earth.

Deflection Coils : In a cathode-ray tube, coils used to bend an electron beam a desired amount.

Deflection Plates : Two pairs of parallel electrodes, one pair set forward of the other and at right angles to each other, parallel to the axis of the electron stream within an electrostatic cathode-ray tube.

Degeneration : The process whereby a part of the output signal of an amplifying device is returned to its input circuit in such a manner that it tends to cancel part of the input.

Degenerative Feedback : Feedback in which the feedback signal is out of phase with the input signal.

Degree-of-Freedom : The number of axes about which a gyro is free to precess.

Deionization Potential : The potential at which ionization of the gas within a gas-filled tube ceases and conduction stops; also referred to as extinction potential.

Deionization Time : In a spark gap, the time required for ionized gas to return to its neutral state after the spark is removed.

Delay Equalizer : A device that adds delay to only certain frequencies in a circuit to reduce delay distortion.

Delay Line : A transmission line, or equivalent passive component, used to delay a signal. A component or device either active or passive used to delay a signal.

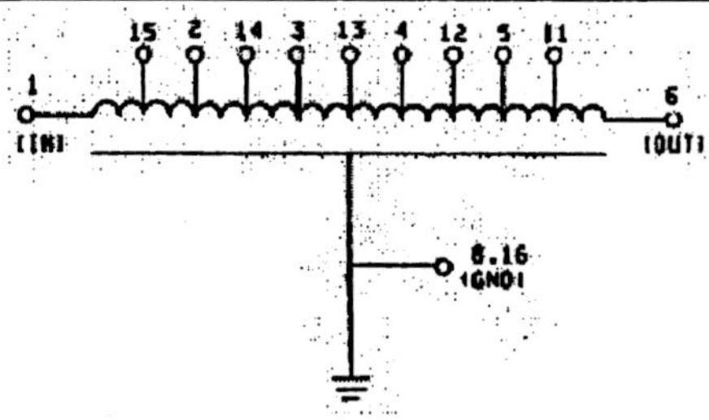

Tapped Delay Line

Delayed Sweep **:** In an oscilloscope when a horizontal sweep is delayed by some amount of time after a trigger is received.

Delimiter **:** A character used to indicate the beginning and end of a character string, i.e., a symbol stream, such as words, groups of words, or frames. A flag that separates and organizes items of data.

Delta **:** A three-phase connection in which windings are connected end-to-end, forming a closed loop that resembles the Greek letter delta. A separate phase wire is then connected to each of the three junctions.

Demodulation **:** The removal of intelligence from a transmission medium. The recovery, from a modulated carrier, of a signal having substantially the same characteristics as the original modulating signal.

Demodulator **:** A circuit used in servo-systems to convert an ac signal to a dc signal. The magnitude of the dc output is determined by the magnitude of the ac input signal, and its polarity is determined by whether the ac input signal is in or out of phase with the ac reference voltage.

Demorgan's Theorem **:** A theorem which states that the inversion of a series of AND applications is equal to the same series of inverted OR applications, or the inversion of a series of OR applications is equal to the same series of inverted AND applications.

Denitrification : The anoxic biological conversion of nitrate to nitrogen gas. It occurs naturally in surface waters low in oxygen, and it can be engineered in wastewater treatment systems.

Density : The compactness of a substance. Mass per unit volume.

Deoxygenation : The consumption of oxygen by the different aquatic organisms as they oxidised materials in the aquatic environment.

Depletion Region : The region in a semiconductor where essentially all free electrons and holes have been swept out by the electrostatic field which exists there.

Derating : The process of reducing a components power dissipation with increasing temperature to insure that the component remains at the same temperature. Derating applies to any device that allows current flow, Active, Passive or Mechanical.

Design for Testability : Design For Testability (or Design for Test, or DFT) refers to design techniques that make products easier to test. Examples include the addition of test points, parametric measurement devices, self-test diagnotics, test modes, and scan design.

Deterministic Jitter : Reproducible jitter within a given system, under controlled conditions. Also known as bounded jitter.

Detonation : Spontaneous and violent ignition of part of the charge, usually in the end gas ahead of the flame, capable of promoting pre-ignition and causing serious engine damage. Compression ratio, fuel quality, ignition advance and intake air temperature are the main factors involved in promoting detonation.

Devitrification : The process in which a glass (noncrystalline or vitreous solid) transforms to a crystalline solid

Diamagnetism : A weak form of induced or nonpermanent magnetism for which the magnetic susceptibility is negative.

Dielectric : A material which acts as an electrical insulator.

Dielectric Constant : The ratio of the permittivity of a medium to that of a vacuum.

Dielectric Losses : Power losses due to the conductance of dielectric materials.

Dielectric Strength : The magnitude of an electric field necessary to cause significant current passage through a dielectric material.

Difference Frequency : Beat Frequency. The difference between the oscillator frequency and the unknown audio frequency.

Difference of Potential : A voltage between two points.

Differential Amplifier : A circuit that amplifies the difference between two input signals.

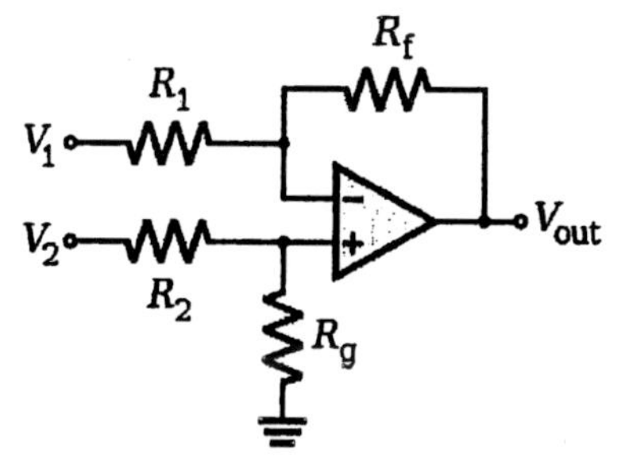

Op Amp Differential Circuit

Differential Pair : A pair of conductors with one line carrying the true version and the other line carrying the inverted version of the signal.

Differential Remote Output Sensing : Uses a Kelvin connection at a remote location to sense the output voltage and better control the voltage at that point.

Differential Signaling : Most electrical signals are single-ended, comprised of a single wire and ground. Differential signals use two wires which are the inverse of each other when one swings positive, the other swings negative in equal magnitude. The receiving circuit looks only at the difference between the two, ignoring any common-mode voltage. This "push-pull" arrangement reduces the impact of electrical interference because external noise will affect both wires equally and the common-mode rejection will ignore the noise.

Diffraction : The bending of waves (as light or RF) when the waves are met with some form of obstruction.

Diffusion : The scattering of reflected light waves from an object, such as white paper. Controlled application of impurity atoms to a semiconductor substrate.

Diffusion Bonding : A solid-state process for joining metals by using heat and pressure to achieve atomic bonding.

Diffusion Coefficient : The constant of proportionality between diffusion flux and the concentration gradient in Fick's first law.

Digit : A symbol that represents one of the nonnegative integers smaller than the radix. For example, in decimal notation a digit is one of the characters from 0 through 9.

Digital Log Pot : Digital logarithmic potentiometer.

Digital Pot : Digital pot or digital potentiometer is a solid-state device that emulates a mechanical potentiometer, it is usually controlled via a simple interface.

Digital System : A system in which data is represented by either a sequence of pulses or by a series of parallel channels each of which has two recognised states, on or off. Binary coded data can be stored or manipulated very rapidly often via an eight channel (eight bit) system capable of representing any number from 0 to 255. Early engine management systems employed 8 bit microprocessors but 16 and 32 bit devices are now common to provide the power necessary for complicated self-diagnostic systems. Input data is usually in analogue form which is converted to digital for processing.

Digitise : Conversion of an analog signal into a digital signal carrying equivalent information.

Diode : A two-terminal device that rectifies signals (passes current in only one direction). Most commonly, a semiconductor consisting of a P-N junction, but dioides can also be realised using vacuum tube, point-contact, metal-semiconductor junction (Schottky), and other technologies.

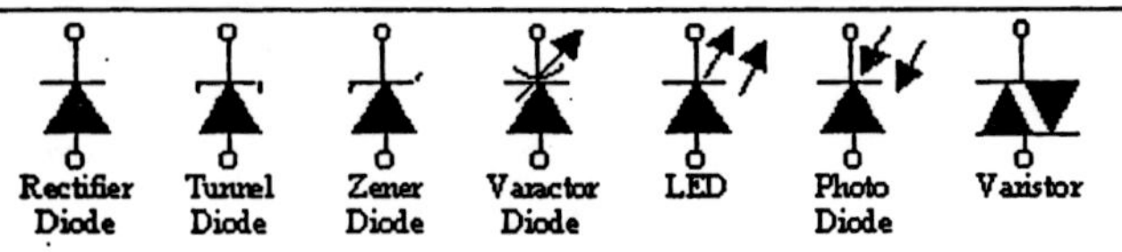

Diode Schematic Symbols

DIP (Dual Inline Package) : DIP is an integrated circuit package with two rows of pins. PDIP (Plastic Dual Inline Package) is a DIP package with a molded plastic body. CDIP (Ceramic Dual Inline Package) is a DIP package with a ceramic body.

Dipole (electric) : A pair of equal yet opposite electrical charges that are separated by a small distance

Discrete Settling : Settling in which individual particles settle independently, neither agglomerating or interfering with the settling of the other particles present. This occurs in waters with a low concentration of particles.

Dislocation : A linear crystalline defect around which there is an atomic misalignment.

Dispersion : A stable mixture of particles suspended in a fluid medium.

Display Cases : Once complete it is a good idea to finish the model and keep it clean with a display case. There are a number of different ways in which a display case may be constructed.

Dissolved Oxygen (DO) : The amount of molecular oxygen dissolved in water.

Distortion : In systems that handle electrical signals, distortion is a generally unwanted change in the signal. Not all signal alterations are considered distortion. For instance, a uniform delay or a linear attenuation or amplification would generally not be considered distortion.

Dithering : A common technique to improve digitising when quantisation noise (quantisation error/noise) can no longer be treated as random. A small amount of random noise is added to the analog input signal. This added noise causes the digital output to randomly toggle between two adjacent codes, thereby avoiding thresholding effect.

Diversity : In radio systems, diversity is a method of improving the reliability and capacity by using multiple communication channels to carry each signal.

DMA (Direct Memory Access) : A scheme which reads or writes data directly to memory, bypassing the processor and the processor bus.

DML (Data Manipulation Language) : A language that allows data to be manipulated in a database. In SQL, commands such as DELETE and INSERT are DML commands.

DMM (Digital Multimeter) : Measuring instrument or VOM (e.g. voltage, resistance, current) with a digital display.

DNL (Differential Nonlinearity) : A specification that appears in data-converter datasheets. In an ideal D/A converter, incrementing the digital code by 1 changes the output voltage by an amount that does not vary across the device's permitted range. Similarly, in an A/D, the digital value ramps smoothly as the input is linearly swept across its entire range. DNL measures the deviation from the ideal. An ideal converter has the code exactly the same size, and a DNL of 0 (zero).

DOCSIS (Data Over Cable Service Interface Specification) : A standard for delivering data over cable TV systems, typically for subscriber Internet access services.

Doping : The intentional alloying of semiconducting materials with controlled concentrations of donor or acceptor impurities.

Down Converters : A device which provides frequency conversion to a lower frequency, e.g. in digital broadcast satellite applications.

Drain : One of the three terminals that comprise a FET. A voltage on the gate controls the current flow between the source and drain.

DRAM (Dynamic RAM) : Random-Access Memory that uses a continuous clock. Unlike SRAM, when DRAM is no longer clockedl its data is lost.

Drawing : A deformation technique used to fabricate metal wire and tubing. Deformation is accomplished by pulling the material through a die by means of a tensile force applied on the exit side.

Drill Bits : Normally made of hardened or high speed steel with two cutting edges on the conical end with two helical flutes that act to move material away from the cutting edges.

DRL (Daytime Running Lamps) : DRLs are white lights mounted on the front of an automobile. Mandated in many countries, they automatically switch on when the key is turned and are intended for daytime use, to increase the visibility of the automobile. They are typically built with LEDs.

Dry Cell : An electrical cell in which the electrolyte is not a liquid. In most dry cells the electrolyte is in the form of a paste.

Dry Joint : A term used to describe a soldered joint which did not fully adhere at the time of manufacture and which has since developed high resistance, often of an intermittent nature. Rarely obvious from visual inspection.

Dry-Air System : Provides dehumidified air for electronic equipment that is moisture critical.

Drypack : Drypack is a method for packing integrated circuits in a moisture-free environment. The device is baked and immediately sealed in a vacuum-sealed bag. This process is reserved for package types which are especially susceptible to moisture intrusion.

DSLAM (Digital Subscriber Line Access Multiplexer) : A device which takes a number of ADSL subscriber lines and concentrates these to a single ATM line.

DSSS (Direct-Sequence Spread Spectrum) : A transmission technology used in WLAN (wireless LAN) transmissions where a data signal at the sending station is combined with a higher data-rate bit sequence, or chipping code, that divides the user data according to a spreading ratio.

D-Sub : Short for D Sub-miniature. A style of connector with the mating end shaped like the letter 'D'.

DTMF (Dual Tone Multiple Frequency) : DTMF is a signaling method developed by Bell Labs for sending telephone dialing information over the same analog, voice-quality phones lines that carry voice. Each digit is encoded as the sum of two sine wave bursts, of different frequencies. The two-tone method was chosen because it can be reliably distinguished from voice and normal phone conversations are highly unlikely to falsely trigger the DTMF receiver.

Dual In-Line Package (DIP) : IC package having two parallel rows of performed leads.

Dual Mode : Two modes of operation. Examples: In power circuits, the IC can deliver either a fixed 5V or an adjustable 1.3V to 16V source. In cellular phones, the IC operates in FM or CDMA mode, AMPS or TDMA mode, etc.

Dual Phase Controller : Switching regulator that employs dual-phase technique to reduce output noise and boost output current capability.

Dual-Band : Dual-band refers to the capability of GSM network infrastructure and handsets to operate across two frequency bands.

Dual-Gate Mosfet : A two-gate MOSFET in which either gate can control the conductor independently, a fact which makes this MOSFET very versatile.

Dual-Modulus Prescaler : A Dual-Modulus Prescaler (DMP) is an important circuit block used in frequency synthesisers to divide the high-frequency signal from the voltage controlled oscillator (VCO) to a low-frequency signal by a predetermined divide ratio, either (N+1) or N, which is controlled by a swallow counter. This low-frequency signal is then further divided by the main counter to the desired channel-spacing frequency which is then fed to the phase detector to form the closed feedback loop in frequency synthesisers.

Duct : In interfacility cabling, a conduit, which may be direct-earth buried or encased in concrete, used to enclose communications or power cables.

Ductile : The propagation of radio waves within an atmospheric duct. Easily drawn out (as to form filaments or wires).

Ductility : A measure of a material's ability to undergo appreciable plastic deformation before fracture.

Ducting : Trapping of an RF wave between two layers of the earth's atmosphere or between an atmospheric layer and the earth.

Dummy Load : A dissipative but non-radiating device that has the impedance characteristics of an antenna or transmission line.

Dump : An illegal and uncontrolled area where wastes have been placed on or in the ground.

Duplex Circuit : A circuit that permits transmission in both directions.

Duplexer : A radar device that switches the antenna from the transmitter to the receiver and vice versa.

Duty Cycle : In a transmitter, ratio of time on to time off.

DWDM (Dense Wave Division Multiplexing) : The technology by which the frequencies of light carried on a single optical fiber are subdivided into discrete wavelengths, allowing for the greater transmission of data.

Dynamic Microphone : A device in which sound waves move a coil of fine wire that is mounted on the back of a diaphragm and located in the magnetic field of a permanent magnet.

Dynamic Range : The range, in dB, between the noise floor of a device and its defined maximum output level.

E

E1 : Wide-area, digital transmission scheme, used predominantly in Europe, that carries data at a rate of 2.048Mbps. E1 lines can be leased for private use from common carriers.

E2 : A line that carries four multiplexed E1 signals with a data rate of 8.448Mbps.

E3 : Wide-area, digital transmission scheme used predominantly in Europe that carries data at a rate of 34.368Mbps. E3 lines can be leased for private use from common carriers.

EAM (Electro-Absorption Modulators) : Chip-level modulation devices often integrated into hybrid transponder devices, alongside lasers.

Earphone : A receiver placed near the ear. A transducer.

Earth Ground : The connection to earth. Ground. The most common connection to earth is via a water pipe. The graphic shows how to connect a Printed Circuit Board [PCB] ground to chassis ground. In many cases chassis ground will also connect to earth ground.

Echo : The reflection of the original sound wave as it bounces off a distant surface. The RF signal reflected back from a radar target.

Echo Box : A resonant cavity device that is used to check the overall performance of a radar system. It receives a portion of the transmitted pulse and retransmits it back to the receiver as a slowly decaying transient.

ECL : Emitter Coupled Logic, a type of digital logic using transistors that do not go into saturation, allowing the devices to operate faster.

Eclipse : A condition in which the satellite is not in view or in direct line of sight with the sun. This happens when the earth is between them.

EconoReset : The simplest form of microprocessor supervisory circuit, it monitors the power supply for the microprocessor and provides only a power-on reset function.

EconOscillator : Low-cost, surface-mount, CMOS oscillator family from Dallas Semiconductor. EconOscillators replace crystal-based oscillators. They need no external crystals or timing components.

E-Core : E-Cores expose the winding to so heat does not get trapped inside and also makes it easier to bring out connections from several windings.

Eddy Current Loss : Losses caused by random current flowing in the core of a transformer. Power is lost in the form of heat.

Eddy Current : Induced circulating currents in a conducting material that are caused by a varying magnetic field.

EDGE (Enhanced Data Rates for GSM Evolution) : An enhanced modulation technique designed to increase network capacity and data rates in GSM networks. EDGE should provide data rates up to 384Kbps.

Edge Triggered Flip Flop : A type of Flip Flop [FF] that requires either a rising edge or falling edge of a clock edge to force the output to change to the level of the input signal [D].

Edison Cap : Also called an Edison Base. A type of screw cap used with light bulbs.

Edison Effect : Also called Richardson Effect. The phenomenon wherein electrons emitted from a heated element within a vacuum tube will flow to a second element that is connected to a positive potential.

Effluent Based Standards : Standards which set concentration or mass per time limits on the effluent being discharged to a receiving water.

Effluent : The fluid exiting a system, process, tank, etc.An effluent from one process can be an influent to another process.

EIA (Electronic Industries Alliance) : Among other things, the EIA sponsors electrical and electronic standards.

Elastomer : A polymeric material that may experience large and reversible elastic deformations.

Electric Motors : There are a number of different types of electric motor.

Electrical Screen : A metal shield which isolates a device from external fields.

Electromagnetic Compatibility : The capability of different electrical systems to coexist in the electromagnetic environment without causing or being subjected to interference.

Electron Beam Welding : A technique for joining materials in which components to be welded are heated by a concentrated beam of high-velocity electrons in vacuo.

Electronegativity : The potential of an atom to attract electrons when the atom is bonded in a compound. The scale is 0 to 4 with 0 being the most electropositive (low attraction) and 4 being the most electronegative (high attraction).

Electronic Control Unit (ECU) : General term for the main electronic circuit module which controls injector pulses and/ or ignition timing according to various sensor inputs.

Electrostatic Pprecipitator : A device which uses an electric field to trap particulate pollutants.

Element : A substance, in chemistry, that cannot be divided into simpler substances by any means ordinarily available. A constituent of a device that contributes directly to its operation (e.g., chip resistor, capacitor, diode, transistor, integrated circuit, surface acoustic wave (SAW), substrate, package, etc., incorporated into a device), is an element of the device. A part of an antenna that can be either an active radiator or a parasitic radiator.

Elementary Reaction : A reaction in which the rate expression corresponds to the stoichiometric equation.

Embedded System : A system in which the computer (generally a microcontroller or microprocessor) is included as an integral part of the system. Often, the computer is relatively invisible to the user, without obvious applications, files, or operating systems. Examples of products with invisible embedded systems are the controller that runs a microwave oven or the engine control system of a modern automobile.

EMC (Electromagnetic Compatibility) : The ability of electronic equipment to be a "good electromagnetic neighbor." It neither causes, nor is susceptible to, electromagnetic interference (within the limits of applicable standards).

Emergency Power : Temporary source of limited electrical power used upon the loss of the normal power source.

EMF (Electromotive Force) : The force that causes electricity to flow between two points with different electrical charges or when there is a difference of potential between the two points. The unit of measurement is volts.

EMI (Electromagnetic Interference) : Unwanted noise from electromagnetic radiation.

Emitter : The element in a transistor that emits current carriers (electrons or holes).

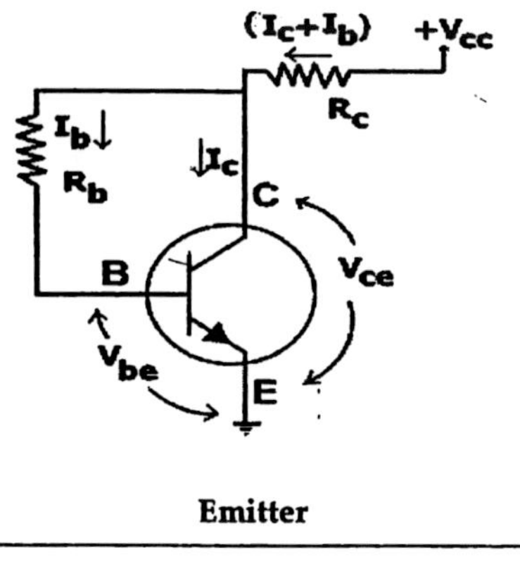

Emitter

Emitter-Injection Modulator **:** The transistor equivalent of the cathode modulator. The gain is varied by changing the voltage on the emitter.

Enamel **:** A synthetic compound of cellulose acetate (wood pulp and magnesium). Used to insulate wire in meters, relays, and motor windings.

Encapsulated **:** Imbedded in solid material or enclosed in glass or metal.

Encode **:** To use a code to represent individual characters in a message.

Encoder **:** A device used to encode a signal or character from one format to another.

End of Life **:** A part or device is no longer being manufactured, however there may or may not be inventory still in stock or carried by a distributor..

End Point **:** Behavior of the device at the limit of temperature or voltage.

Energized **:** Electrically connected to a source of potential difference, or electrically charged so as to have a potential significantly different from that of earth in the vicinity.

Energy **:** The ability or capacity to do work.

Engine Control Module (ECM) **:** A term that seems to have taken over from ECU, having much the same meaning.

Engine Management System **:** A complex type of microprocessor based ECU which controls both fuelling and ignition.

Engine Specific Volume **:** This is the engine bulk volume divided by power.

Engine Specific Weight **:** This is the engine weight divided by power.

Engineering Ceramics **:** Technical ceramics for structural applications.

ENOB (Effective Number of Bits) **:** An indication of the quality of an analog-to-digital converter (ADC). The measurement is related to the test frequency and the signal-to-noise ratio.

Ephemeris : A table showing the pre-calculated position of a satellite at any given time.

Ep-Ip Curve : The characteristic curve of an electron tube used to graphically depict the relationship between plate voltage (Ep) and plate current (Ip).

Epitaxial Process : A method of depositing a thin, uniformly doped crystalline region (layer) on a substrate.

Equatorial Orbit : An orbit that occurs when the plane of a satellite coincides with the plane of the earth at the equator.

Equipment : A general term characterizing the broad category of electronic items.

Equivalent : The mass of the compound which will produce one mole of available reacting substance. Thus, for an acid, this would be the mass of acid which will produce one mole of : H+, for a base, one mole of OH-.

Equivalent Resistance (Req) : A resistance that represents the total ohmic values of a circuit component or group of circuit components. Usually drawn as a single resistor in a simplified circuit.

Erasable Programmable Read Only Memory (EPROM) : A type of microchip which can be programmed at manufacture with the fuel, ignition, and other functions of the ECU. A window (which must be covered over to keep out daylight) permits erasure by exposure to UV light, after which the device can be reprogrammed using special equipment. An alternative (PROM) is the same thing but cannot be erased. Electrically erasable type (EEPROM) also available from which data can be selectively erased.

ESD *(Electrostatic Discharge)* : Release of stored static electricity. The potentially damaging discharge of many thousands of volts that occurs when an electronic device is touched by a charged body.

ESD *Protection* : Devices added to input and output pins on an IC to protect the internal circuitry from the damaging effect of electrostatic discharge.

ESD Strap : A device to ground and discharge static electricity from a person.

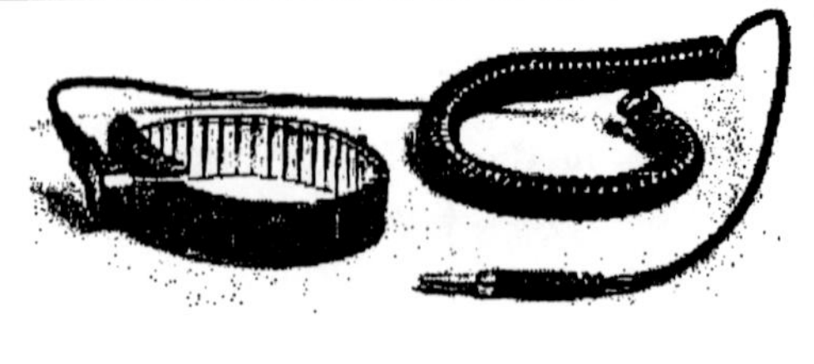

ESD Wrist Strap

ESF (Extended Superframe) : A DS1 framing format in which 24 DS0 times lots, plus a coded framing bit are organised into a frame which is repeated 24 times to form a superframe.

ESR (Effective Series Resistance) : ESR is the resistive component of a capacitor's equivalent circuit. A capacitor can be modeled as an ideal capacitor in series with a resistor and an inductor. The resistor's value is the ESR.

Ethernet : A family of network protocols based on asynchronous frames. The Ethernet framing structure provides a flexible payload container with basic addressing and error detection mechanisms.

Ethernet MAC Address (MAC Address): A unique 6-byte address that identifies an Ethernet device on the network. In an Ethernet packet, a MAC address is used twice, first to identify the packet's destination, and second to identify the packet's sender or source. An Ethernet MAC address is normally specified in hexadecimal, using dashes to separate bytes. For example, 08h-00h-28h-32h-17h-42h. The first three bytes normally designate the manufacturer of the device. However, when the first byte of the address is odd (LSB is 1), the address is a group address (broadcast or multicast). The second bit specifies whether the address is globally or locally administrated.

Ethers : An organic compound which has two hydrocarbon groups bound by an interior oxygen atom. The general formula is R'-O-R".

E-Transformer : A special form of differential transformer employing an E-shaped core. The secondaries of the transformer are wound on the outer legs of the E, and the primary is wound on the center leg. An output voltage is developed across the secondary coils when its armature is displaced from its neutral position. This device is used as an error detector in servo-systems that have limited load movements.

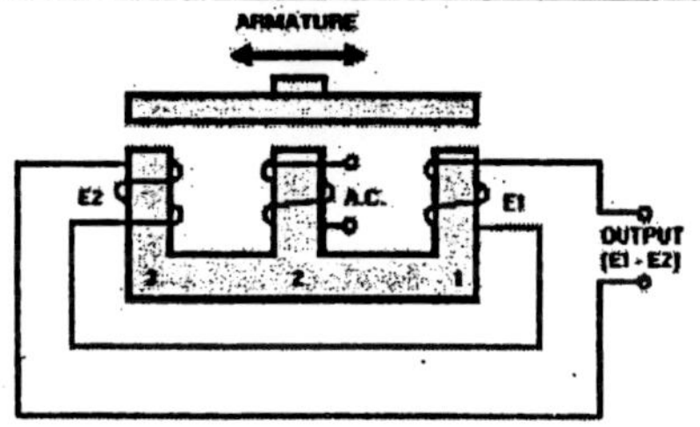

E-Transformer

E-Type T-Junction : A waveguide junction in which the junction arm extends from the main waveguide in the same direction as the E-field in the waveguide.

Eucaryotic Organisms : Organisms which possess a nuclear membrane. This includes all known organisms except viruses and bacteria.

Eutectic Alloy : An alloy that changes directly from a solid to a liquid with no plastic or semi-liquid state.

Eutectic Phase : One of the two phases found in the eutectic structure.

Eutectic Solder : An alloy of 63 percent tin and 37 percent lead. Melts at 361º F.

Evaluation Kit : Evaluation kit is a printed circuit board with an integrated circuit and support components to produce a working circuit for evaluation and development. Most Evaluation Kits are fully assembled and tested.

Evaporation : The conversion of liquid water to water vapour. It occurs on the surface of water bodies such as lakes and rivers and immediately after precipitation events in small depressions and other storage areas.

Evaporative Loss Emission System : A system normally based on a charcoal filled canister which traps evaporative vapours from the fuel tank. A canister purge system uses engine manifold vacuum in certain operating circumstances to draw air through the charcoal so that the vapour is passed into the engine to be burnt as fuel. This is often controlled so that the slight richening effect provides a check function for the Lambda feedback system.

Evapotranspiration : The sum of evaporation and transpiration. Since it is difficult to measure the two terms independently, they are often grouped as one value.

EVM (Error Vector Magnitude) : A measure of the difference between the (ideal) waveform and the measured waveform. The difference is called the error vector, usually referred to with regard to M-ary I/Q modulation schemes like QPSK, and shown on an I/Q "constellation" plot of the demodulated symbols.

EVSYS (Evaluation System) : Evaluation kits that also include an interface board for connecting to a personal computer and Windows-based EVKit software.

Excitation Voltage : The supply voltage required to activate a circuit.

Exciting Current : The current that flows in the primary winding of a transformer when the secondary is open-circuited; it produces a magnetic flux field.

Exclusive OR : A function whose output is a 1 if one and only one of the input variables is a 1.

Exclusive-NOR Gate : A gate that produces a logic 0 output when the inputs are different, but not when they are the same.

Exclusive-OR Gate : A gate that produces a logic 1 output when the inputs are different, but not when they are the same.

Exhaust Catalyst : A chamber in the exhaust system containing, in either matrix or pellet form, certain compounds, often comprised of precious metals, which promote reactions in the exhaust after it has passed out of the engine. Originally used for oxidation only in conjunction with an air pump. Becomes inactive if exposed to lead or certain other compounds like phosporous and sulphur that are now restricted to very small amounts in fuel. Some catalysts perform an effective silencing function.

Exhaust Gas Recirculation (EGR) : A widely used method for reducing exhaust emission of Oxides of Nitrogen (NOx) by feeding a small proportion of spent exhaust gas back into the incoming charge. Has the effect of reducing combustion temperatures. Normally controlled by an EGR valve which limits the function to certain emission critical operating regimes. Can also be provided by extending valve overlap. Some basic parameters describing the exhaust system used with internal combustion engines to increase power and reduce noise.

Exponent : The numeral written in superscript (10^2) which indicates the power to which the base is to be raised.

Exposed Pad : Offered in some packages to improve thermal dissipation or lower the impedance of the ground connection. Normally not electrically isolated, it typically needs to be connected to a ground or power plane, depending on the device.

Expression : A validated series of variables, constants, and functions that can be connected by operating symbols to describe a desired computation.

Externally Excited Meter : A term used to describe meters that get their power from the circuit to which they are connected.

Extractor Exhaust : An exhaust system using specific pipe dimensions to harness pulses generated as each exhaust valve opens. Positive pressure pulses travelling along the system are reflected from the end as negative pulses which return to augment the induction and scavenging processes in the cylinders. All racing systems are of this type and can contribute as much as 30% to the power output. Ram induction is a similar action taking place in the induction system.

Extremely High Frequency : The band of frequencies from 30 gigahertz to 300 gigahertz.

Extremely Low Frequency : The band of frequencies up to 300 hertz.

Extrinsic : A semiconductor in which impurities have been added to create certain charge carrier concentrations.

Extrinsic Semiconductor : A semi-conducting material for rich the electrical behavior is determined by impurities.

Extrusion : A forming technique whereby a material is forced, by compression, through a die orifice.

Eye Pattern : An oscilloscope display in which a pseudorandom digital data signal from a receiver is repetitively sampled and applied to the vertical input, while the data rate is used to trigger the horizontal sweep.

F

F **: (1) Farad(s): Unit of capacitance. (2) F in lower case is the standard abbreviation for femto, a metric prefix for 10 to the -15.**

Face-centered Cubic (FCC) **: A crystal structure found in some of the common elemental metals. Within the cubic unit cell, atoms are located at all corner and face-centered positions.**

Factor **: Any of the elements, quantities, or symbols that, when multiplied together, form a product.**

Facultative **: A group of microorganisms which prefer or preferentially use molecular oxygen when available, but are capable of suing other pathways for energy and synthesis if molecular oxygen is not available.**

Failsafe **: Built-in safety characteristics of a unit or system so that unit or system failure or a loss of control power will not result in an unsafe condition. The design feature of a part, unit or equipment which allows the item to fail only into a non-hazardous mode.**

Fan **: A component designed to move air. A set of axial fans are shown to the right, plastic box type.**

Fans

Fan Controller - Linear **:** An integrated circuit that varies the speed and airflow of a cooling fan using a variable voltage in response to temperature or system commands.

Fan Controller - PWM **:** An integrated circuit that varies the speed and airflow of a cooling fan using a pulse-width-modulated (PWM) voltage in response to temperature or system commands.

Fanout **:** The number of IC gates an Integrated circuit can drive is determined by the current it can source and sink.

Far End Crosstalk Ratio **:** Crosstalk measured at the far end of a transmission line.

Far Field **:** The electromagnetic field which exists at a distance of several wavelengths from the source of radiation.

Farad **:** The basic unit of capacitance. A capacitor has a capacitance of 1 farad when a voltage potential of 1 volt across it produces a charge of 1 coulomb.

Faraday Rotation **:** The rotation of the plane of polarisation of electromagnetic energy when it passes a substance influenced by a magnetic field that has a component in the direction of propagation.

Faraday Shield **:** A common conductor connection at one end to provide electrostatic shielding without effecting electromagnetic waves. The common point connection is normally grounded.

Fast Bus **:** IEEE Std 960-1993, Fast bus Defined the Mechanical, Electrical, and Protocol for a board-to-backplane interface. This interface is obsolete.

Fast-Time-Constant Circuit **:** Differentiated circuit in the first video amplifier that allows only the leading edges of target returns, no matter how small or large, to be used.

Fatigue **:** Failure, at relatively low stress levels, of structures that are subjected to fluctuating and cyclic stresses.

Fault Blanking **:** A function that ignores a fault for a predetermined period. This is done to eliminate nuisance fault indication.

Fault Indicator : A component or device that indicates a failure or defect in the equipment being monitored either by audio or visual indication.

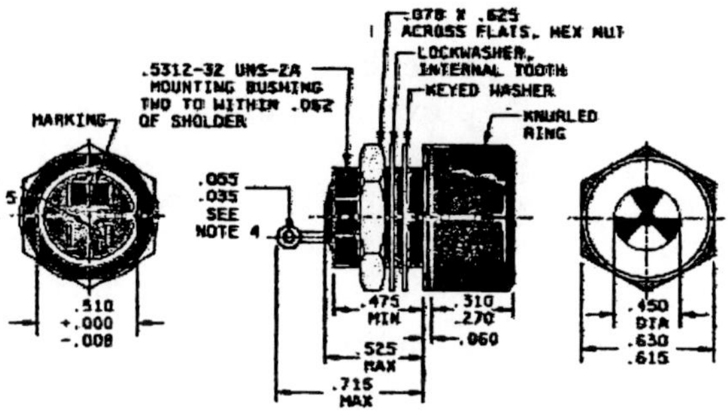

Fault Indicator

Fault Tolerant : Will tolerate excessive voltage during a fault condition.

FDDI (Fiber Distributed Data Interface) : A standard for transmitting data on optical fiber cables at a rate of around 100,000,000 bits-per-second (10 times as fast as 10 Base-T Ethernet; about twice as fast as T-3).

FDL (Facility Data Link) : Embedded communications channel in ESF DS1 framing. Used to convey both bit-oriented and message-oriented signals.

FDM : A method for carrying multiple channels of information on one channel by dividing the available bandwidth among the channels.

FEC (Forward Error Correction) : A technique for detecting and correcting errors from imperfect transmission by adding a small number of extra bits. FEC allows optical transmission over longer distances by correcting errors that can happen as the signal-to-noise ratio decreases with distance.

Feed-Thru Capacitor : A bulk-head mount filter used to pass signals into an enclosure. The most common filter element is a shunt capacitor.

Femto Base Station **:** A femto base station (also called an Access Point Base Station, femtocell, femtobasestation or femto base station) is an in-home base station. Like a standard base station, it connects cell phone voice and data to the cell phone network, but it serves a smaller area (the home). A femto base station benefits the service provider because it off loads cell tower traffic. Subscribers benefit from superior signal strength, due to the proximity of the unit — especially where a cellular signal is weak or not available. Femto base stations augment the normal network and replicates the usual telecommunications infrastructure. Connection to the cell phone network is provided by VoIP over the Internet.

Femto Base Station **:** A femto base station (also called an Access Point Base Station, femtocell, femtobasestation or femto base station) is an in-home base station. Like a standard base station, it connects cell phone voice and data to the cell phone network, but it serves a smaller area (the home). A femto base station benefits the service provider because it off loads cell tower traffic. Subscribers benefit from superior signal strength, due to the proximity of the unit — especially where a cellular signal is weak or not available.

FEP **:** Fluorinated Ethylene Propylene, A synthetic type of insulation.

Fermentation **:** Energy production without the benefit of oxygen as a terminal electron acceptor, i.e. oxidation in which the net effect is one organic compound oxidising another.

Fermi Energy **:** For a metal, the energy corresponding to the highest filled electron state in the valence bond at 0 K.

Ferrite (iron) **:** Body-centered cubic iron. Also, iron and steel alloys that have the BCC crystal structure.

Ferrite Switch **:** A ferrite device that blocks the flow of energy through a wave guide by rotating the electric field 90 degrees. The rotated energy is then reflected or absorbed.

Ferrite **:** A powdered and compressed ferric oxide material that has both magnetic properties and light resistance to current flow.

Ferroelectric : A dielectric material that may exhibit polarisation in the absence of an electric field.

Ferromagnetic Material : A highly magnetic material, such as iron, cobalt, nickel, or their alloys.

Ferromagnetism : Permanent and large magnetisations found in some metals (e.g., Fe, Ni, and Co), which result from the parallel alignments of neighbouring magnetic moments.

Ferrule : A short metal tube used to make crimp connections to shielded or coaxial cables. The cylindrical metallic ends of a cartridge fuse or wire.

FET (Field-Effect Transistor) : A transistor in which the voltage on one terminal (the gate) creates a field that allows or disallows conduction between the other two terminals (the source and drain). There are three varieties: The JFET (Junction Field-Effect Transistor); the MOSFET (Metal-Oxide-Semiconductor Field-Effect Transistor); and the MESFET (Metal-Semiconductor Field-Effect Transistor). The FET is one of two major kinds of transistor, the other being the Bipolar Junction Transistor. A Fourier transform (FT) converts a signal from the time domain (signal strength as a function of time) to the frequency domain (signal strength as a function of frequency). It shows the signal's spectral content, divided into discrete bins (frequency bands).

FHSS (Frequency Hopping Spread Spectrum) : A transmission technology in which the data signal is modulated by a narrowband carrier signal which changes frequency ("hops") over a wide band of frequencies. The hopping seems random but is prescribed by an algorithm known to the receiving system.

Fiber : Any material that has been drawn into a cylinder with a length-to-diameter ratio greater than about ten.

Fiber Optic Cable : A fiber, multiple fiber or fiber bundle in a cable structure fabricated to meet optical mechanical and environmental specifications.

Fiber Optic Connector : A fiber optic component normally assembled onto a cable and attached to a piece of apparatus for the purpose of providing interconnecting/disconnecting of fiber optic cables.

Fiber Optics **:** Conductors or optical waveguides that readily pass light.

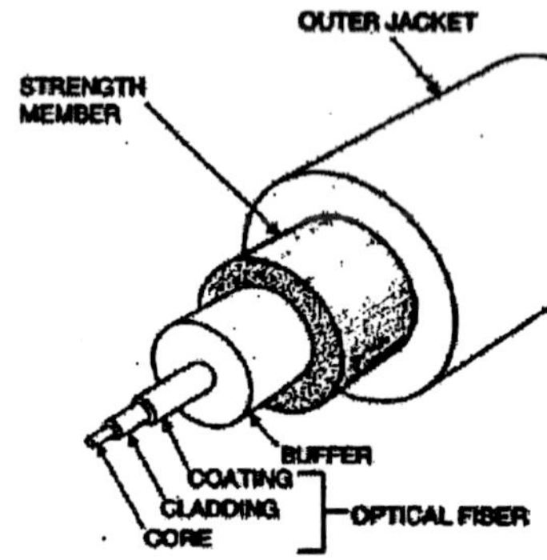

Optical Fiber

Fibre Channel **:** A highly-reliable, gigabit interconnect technology that allows concurrent communications among workstations, mainframes, servers, data storage systems, and other peripherals using SCSI and IP protocols. It provides interconnect systems for multiple topologies that can scale to a total system bandwidth on the order of a terabit per second.

Fibrous Braid **:** An outer covering used to protect a conductor's insulating material. Commonly made from cotton, linen, silk, rayon, or fiberglass. Cable braid is also made off metal and used as a conductor shield or signal conductor.

Fidelity **:** The faithful reproduction of a signal. The accuracy with which a system reproduces a signal at its output that faithfully maintains the essential characteristics of the input signal.

Field **:** The electromagnet which furnishes the magnetic field that interacts with the armature in motors and generators.

Field-Effect Transistor (FET) **:** A transistor consisting of a source, a gate, and a drain. Current flow is controlled by the transverse electric field under the gate. A transistor in which the conduction is due entirely to the flow of majority carriers through a conduction channel controlled by an electric field arising from a voltage applied between the gate and source terminals.

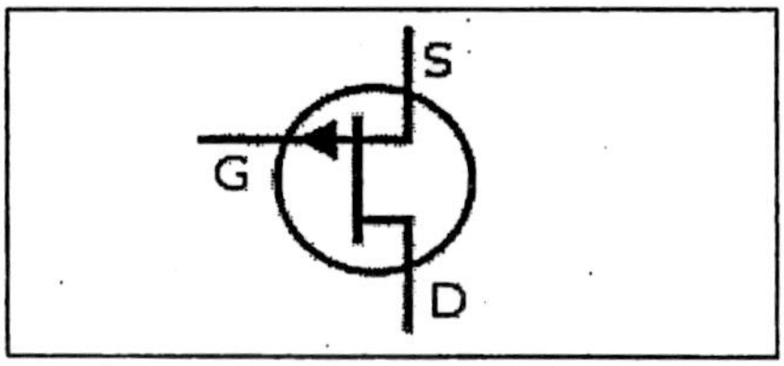

Field Excitation **:** The creation of a steady magnetic field within the field windings by the application of a dc voltage either from the generator itself or from an external source.

Field of Force **:** A term used to describe the total force exerted by an action-at-a-distance phenomenon such as gravity upon matter, electric charges acting upon electric charges, and magnetic forces acting upon other magnets or magnetic materials.

Filament **:** The cathode of a thermionic tube, usually a wire or ribbon, which is heated by current passing through it.

FIFO (First-In First Out) **:** A type of memory that stores data serially, where the first bit read is the first bit that was stored.

Filler **:** An inert foreign substance added to a matrix to improve or modify its properties.

Film ICs **:** Conductive or non-conductive material deposited on a glass or ceramic substrate. Used for passive circuit components, resistors, and capacitors.

Filter **:** A selective network of resistors, capacitors, and inductors that offers comparatively little opposition to certain frequencies, while blocking or attenuating other frequencies. In electronics, a device that transmits only part of the incident energy and may thereby change the spectral distribution of energy: (a) high-pass filters transmit energy above a certain frequency; (b) low-pass filters transmit energy below a certain frequency; (c) bandpass filters transmit energy of a certain bandwidth; (d) band-stop filters transmit energy outside a specific frequency band.

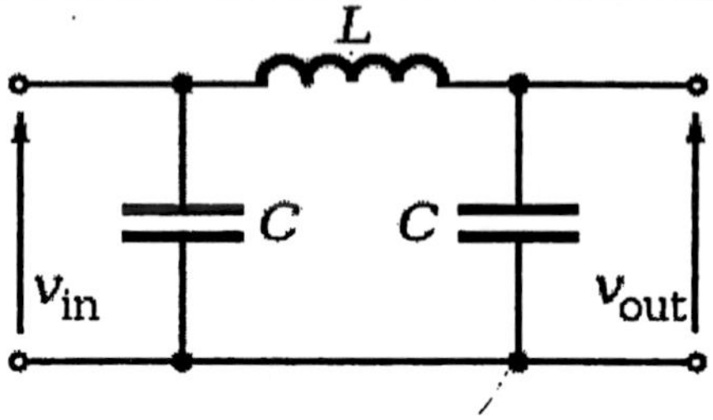

Low Pass PI Filter

Fir **:** A light coloured and soft wood. Needs to be dried very well before use.

FireWire **:** Apple Computer trademarked name for the IEEE 1394 serial interface standard: A highspeed interface between computers and peripherals such as external disk drives, cameras, and camcorders.

Firing **:** A high-temperature heat treatment that increases the density and strength of a ceramic piece.

Fixed Solids **:** FS are the solids that do not volatilise at 550°C.

Fixed Suspended Solids (FSS) **:** FSS is the matter remaining from the suspended solids analysis which will not burn at 550°C. It represents the non-filterable inorganic residue in a sample.

Flame Engine **:** These engines have been variously called Atmospheric Engines, Vacuum Engines and Flame Suckers. This particular example is an R-M-C model engine.

Flash ADCs **:** An analog-to-digital converter that uses a series of comparators with different threshold voltages to convert an analog signal to a digital output.

Flash Point **:** The lowest temperature at which sufficient vapor is produced to cause combustion if an ignition source is present.

Flocculant Settling : Settling in which particle concentrations are sufficiently high that particle agglomeration occurs. This results in a reduction in the number of particles and an increase in average particle mass. : As agglomeration occurs higher settling velocities result.

Fluidisation : The suspension of particles by sufficient upward velocity of the fluid.During fluidisation the gravity force is overcome by a combination of buoyancy and fluid friction.

Flux : The movement of a mass past a surface, plane, or boundary. The units are mass per unit area per unit time or [Kg/m^{2}-hour].

FM (Frequency Modulation) : A modulation method in which the carrier frequency changes with the input signal amplitude.

Foldback Current Limit : A circuit which reduces the current limit once the device enters current-limited operation. Commonly seen on RS-422/RS-485 drivers and some power circuits.

Force-Sense : Measurement technique in which a voltage (or current) is forced at a remote point in a circuit; then the resulting current (or voltage) is measured (sensed).

Forging : Mechanical forming of a metal or alloy by heating and hammering.

Forward Converter : A power-supply switching circuit that transfers energy to the transformer secondary when the switching transistor is on.

FPGA (Field Programmable Gate Array) : A family of general-purpose logic devices that can be configured by the end user to perform many, different, complex logic functions. It is often used for prototyping logic hardware.

Fracture Toughness : Critical value of the stress intensity factor for which crack extensions occurs.

Frame Relay : A highspeed, packet-switched data communications service similar to X.25. Frame relay is a leading contender for LAN-to-LAN interconnect services, and is well suited to the burst-intensive demands of LAN environments.

o

Framer : A device used to align/synchronise to an embedded framing pattern in a serial bit stream. Once synchronised and data fields are properly aligned, overhead bits for alarms, performance monitoring, embedded signaling, etc. May be extracted and processed.

Free Energy : A thermodynamic quantity that is a function of both the internal energy and entropy of a system.

Frenkel Defect : In an ionic solid, a cation-vacancy and cation-interstitial pair.

Frequency (f) : The number of complete cycles per second existing in any form of wave motion, such as the number of cycles per second of an alternating current. The rate at which the vector that generates a sine wave rotates.

Frequency Bin : The frequency range and resolution on the frequency axis of a spectrum graph depends on the sampling rate and the size of the data record (the number of acquisition points). The number of frequency points or lines or bands in the power spectrum is NRECORD/2, where NRECORD is the number of signal points captured in the time domain. The first frequency line in the power spectrum always represents DC (frequency=0). The last frequency line can be found at fSAMPLE/2 - fSAMPLE/NRECORD. Frequency lines are spaced at even intervals of fSAMPLE/NRECORD, commonly referred to as frequency or FFT bins or frequency bins.

Frequency Compensation Network : Circuit modification used to improve or broaden the linearity of its frequency response.

Frequency Cutoff : The frequency at which the filter circuit changes from an action of rejecting the unwanted frequencies to an action of passing the desired frequencies. Conversely, the point at which the filter circuit changes from an action in which it passes the desired frequencies to an action in which it rejects the undesired frequencies.

Frequency-Determining Network : A circuit that provides the desired response (maximum or minimum impedance) at a specific frequency.

Frequency Deviation : The amount the frequency varies from the carrier frequency.

Frequency Diversity : In radio systems, Frequency Diversity spreads a signal across multiple channels by transmitting multiple versions of the signal on different frequencies.

Frequency-Division Multiplexing : Multiplexing that transmits and receives the full 360 degrees of each sine wave.

Frequency Meter : A meter used to measure the frequency of an ac signal. An electronic instrument that displays the frequency of an incoming signal. The frequency is determined by comparing the signal to the counter's time base oscillator.

Frequency Mixer : A circuit that accepts two input signals at two different frequencies, and produces an output frequency equal to the difference of the two inputs.

Frequency Modulation (FM) : A signaling method that varies the carrier frequency in proportion to the amplitude of the modulating signal. Angle modulation in which the modulating signal causes the carrier frequency to vary. The amplitude of the modulating signal determines how far the frequency changes, and the frequency of the modulating signal determines how fast the frequency changes.

Frequency Multipliers : Special RF power amplifiers that multiply the input frequency. A circuit that converts an incoming signal to a higher frequency by adding a fixed multiple of cycles or pulses to the signal.

Frequency Response : The measurement of how a circuit or device reacts to different frequencies that are applied to it. The part of a frequency regain than can be sensed by a device. The measure of a servo's ability to respond to various input frequencies.

Frequency-Response Curve : A curve showing the output of an amplifier (or any other device) in terms of voltage or current plotted against frequency with a fixed-amplitude input signal.

Frequency Scanning : Varying the output frequency to achieve electronic scanning.

Frequency-Shift Keying (FSK) : Frequency modulation somewhat similar to continuous-wave (cw) keying in AM transmitters. The carrier is shifted between two differing frequencies by opening and closing a key.

Frequency Stability : Refers to the ability of an oscillator to accurately maintain its operating frequency. The degree to which an oscillating signal produces the same frequency for a specified interval of time.

Frequency Standard : An oscillator used as a reference source for frequency measurements.

Frequency Synthesis : A process that uses hetrodyning and frequency selection to produce a signal.

Frequency Synthesizer : A frequency source of high accuracy. A bank of oscillators in which the outputs can be mixed in various combinations to produce a wide range of frequencies.

FSK (Frequency Shift Keying) : A method of transmitting digital data by shifting the frequency of a carrier signal to represent binary 1s and 0s.

FTTH (Fiber-to-the-home) : A method for broadband data (voice, Internet, multimedia, etc.) Delivery to the home via optical fiber.

FTTN : FTTN is "Fiber-to-the-node." There are two technologies for delivering broadband: Fiber-to-the-node (FTTN) uses fiber to bring data to a node and uses copper to bring the data into the home. Fiber-to-the-home (FTTH) brings fiber all the way into the home.

Fuel Cell : A fuel cell converts energy generated by chemical reaction into electrical energy. Its function is similar to a battery, except that the chemicals generally are supplied from outside the cell.

Fuel Damper : A device very much like a pressure regulator which is sometimes included in either the fuel feed or return pipe to eliminate pulsations or noise.

Fuel Map **:** A three dimensional representation of the relationship between fueling and sensor inputs, normally considered in terms of engine speed and load, although other variables can also be used.

Fuel Pressure Regulator **:** A relief valve which controls fuel pressure to a constant value. Early versions held pressure constant on an absolute basis according to spring pressure, later systems maintain pressure constant in relation to manifold pressure sensed via an interconnecting pipe.

Fuel Rail **:** The metal tube which disperses fuel to the individual injectors.

Full Duplex **:** A channel providing simultaneous transmission in both directions.

Full Load Enrichment **:** Extra fuel provided at full load usually activated by the throttle switch (or potentiometer) as the throttle approaches fully open or from a manifold pressure switch. Full load enrichment is generally of the order 10 to 15% added to the basic fueling.

Full Load Pressure Switch **:** A switch sometimes used in addition or in preference to the throttle switch to detect the approach of full load to activate enrichment. Usually set to function at about 3 ins of mercury depression.

Full-Wave Rectifier **:** A circuit that uses both positive and negative alternations in an alternating current to produce direct current. There are two general circuit configurations; A two diode circuit that requires a center tapped transformer and a four diode approach that does not require a center tapped transformer.

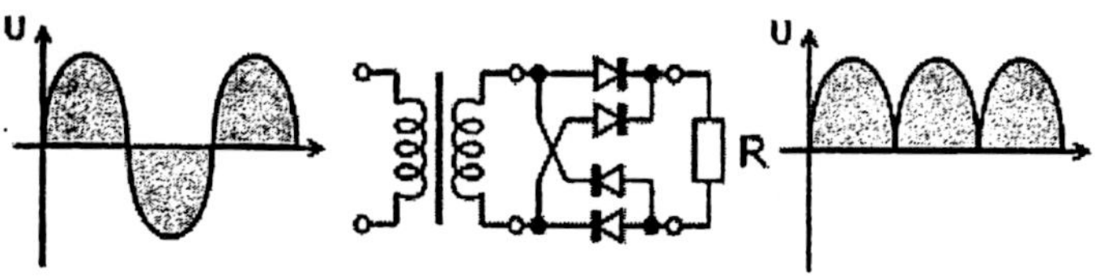

Full-Wave Rectifier Circuit

Full-Wave Voltage Doubler **:** Consists of two half-wave voltage rectifiers and is used to reduce the output ripple amplitude.

Function **:** A specific purpose of an entity; its characteristic action.

Fundamental Frequency **:** The basic frequency or first harmonic frequency.

Fungus-inert material **:** A material which, in all modified states and grades, is not a nutrient to fungi.

Fuse **:** A device that has as its critical component a metal wire or strip that will melt when heated by a prescribed (design) amperage, creating an open in the circuit of which it is a part, thereby protecting the circuit from an over current condition.

Fused Holder **:** A device used to hold a fuse. The shape of the fuse holder used is determined by the style of fuse being used. Two common style of fuse holders include Block Fuse Holders, and Panel Mount Fuse Holders.

Futurebus **:** IEEE-896. Futurebus is a back-plane bus standard that defines both the Physical and Electrical layers for the interface. The follow-on specification was called Futurebus+, which was an upgrade.

G

GaAs (Gallium Arsenide) : A semiconductor material used for optoelectronic products such as LEDs, and for highspeed electronic devices.

GaAs MESFET : A Gallium Arsenide (GaAs) Metal-Semiconductor Field-Effect-Transistor (MESFET) is a transistor built with gallium arsenide semiconductor material. The conducting channel is built using a metal-semiconductor (Schottky) junction.

Gain : The amount of amplification accomplished by an amplifier circuit. For instance, a gain of 2 would mean the output is scaled to twice the amplitude of the input.

Gain Error : The gain error of a data converter indicates how well the slope of an actual transfer function matches the slope of the ideal transfer function. Gain error is usually expressed in LSB or as a percent of full-scale range. Gain error can be calibrated out with hardware or in software. Gain error is the full-scale error minus the offset error.

Gain-Bandwidth Product : The number that results when the gain of a circuit is multiplied by the bandwidth of that circuit. For an operational amplifier, the gain-bandwidth product for one configuration will always equal the gain-bandwidth product for any other configuration of the same amplifier.

Galvanic Isolation : Refers to a design technique that separates signal current from AC power distribution introduced stray noise current.

Galvanometer : A meter used to measure small values of current by electromagnetic or electrodynamic means.

Gamma Correction : The application of a function that transforms brightness or luminance values. Gamma functions are usually nonlinear but monotonic and designed to affect the highlights (whitest values), midtones (grayscale), and shadows (dark areas) separately. Most commonly applied to make a light-emitting device, such as a display, match the human eye's brightness curve. In other terms: A gamma correction function can be used to alter the luminance (light intensity) of a display such that its brightness (the human-perceived values) looks correct.

Gamma : The transistor emitter-to-base current ratio in a common-collector configuration. Gamma Ray; Emissions of a specific frequency produced from subatomic particle interaction.

Ganged Tuning : The process used to tune two or more circuits with a single control.

Gas Stripping : Gas transfer of an undesirable gas from a water stream to the atmosphere.

Gate : As applied to logic circuitry, one of several different types of electronic devices that will provide a particular output when specified input conditions are satisfied. Gate; as in AND Gate, OR Gate, NOR Gate, or NAND Gate. Also, a circuit in which a signal switches another signal on or off.

Gated AGC : Circuit that permits automatic gain control to function only during short time intervals.

Gated-Beam Detector : An FM demodulator that uses a special gated-beam tube to limit, detect, and amplify the received FM signal. Also known as a quadrature detector.

Gating : The process of selecting those portions of a wave that exist during one or more selected time intervals or that have magnitudes between selected limits. Also, the application of a specific waveform to perform electronic switching.

GBIC (Gigabit Interface Converter) : A removable transceiver module permitting Fibre-Channel and Gigabit-Ethernet physical-layer transport.

Gender Changer : A component that changes the Gender of a connector or cable assembly.

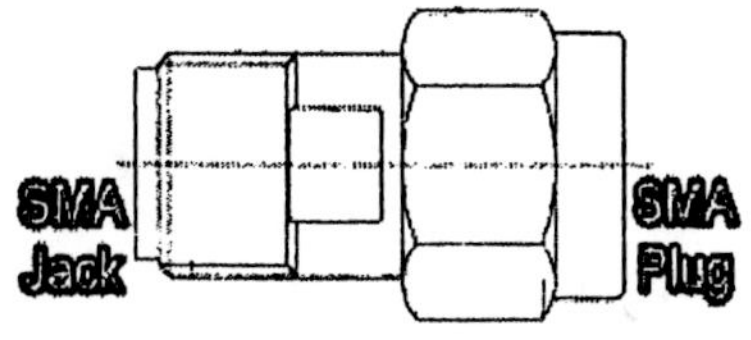

SMA Gender Changer

Generator : A machine that converts mechanical energy to electrical energy by applying the principle of magnetic induction. A machine that produces ac or dc voltage, depending on the original design.

Germanium : Symbol Ge in the periodic table. A grayish-white metal having semiconductor properties, similar to tin.

GFSK ***(Gaussian frequency-shift keying)*** : A type of FSK modulation which uses a Gaussian filter to shape the pulses before they are modulated. This reduces the spectral bandwidth and out-of-band spectrum, to meet adjacent-channel power rejection requirements. Bluetooth uses GFSK.

Gigabit : 1 billion bits-per-second.

Gimbal : A mechanical frame, with two perpendicular intersecting axes of rotation, used to support and furnish a gyro wheel with the necessary freedom to tilt in any direction.

Glass : An inorganic product of fusion which has cooled to a rigid condition without crystallizing.

Glass-ceramic : A fine-grained crystalline material that was formed as a glass and subsequently devitrified (crystallized).

Glitch : General term used to describe an undesirable, momentary pulse or unexpected input or output.

Glitch Immunity : A term used in microprocessor supervisory circuit datasheets to describe the maximum magnitude and duration of a negative-going VCC supply-voltage pulse without causing the reset output to assert.

Global Positioning System (GPS) **:** A US satellite system that lets those on the ground, on the water or in the air determine their position using GPS receivers. Additional GPS data and links.

Global Warming : The long-term warming of the plant due to increases in greenhouse gases which trap reflected light preventing it from exiting to space.

Glow Discharge : Discharge of electricity through a gas in an electron tube.

Glycerine : A raw material used in foodstuffs and printing ink amongst other uses.

GMSK (Gaussian minimum shift keying) : GMSK is a form of frequency shift keying (FSK) used in GSM systems. The tone frequencies are separated by exactly half the bit rate. It has high spectral efficiency.

GPIB (General Purpose Interface Bus) : A standard bus for controlling electronic instruments with a computer. Also called IEEE-488 bus because it is defined by ANSI/IEEE Standards 488-1978, and 488.2-1987. Also called HP-IB, a trademarked term of Hewlett-Packard, which invented the protocol.

GPIO (General Purpose I/O) : A flexible parallel interface that allows a variety of custom connections.

GPRS (General Packet Radio Service) : A radio technology for GSM networks that adds packet-switching protocols and shorter setup time for ISP connections; it offers the possibility to charge by amount of data sent rather than connect time.

GPS (Global Positioning System) **:** A satellite- based navigation system in which two or more signals, received from satellites, are used to determine the receiver's position on the globe.

Gramme-Ring Armature **:** An inefficient type of armature winding in which many of the turns are shielded from the field by its own iron ring.

Graphite **:** The other form of crystalline carbon is diamond. Often used as a lubricant.

Gray Code : A positional binary number notation in which any two numbers whose difference is one is represented by expressions that are the same except in one place.

Greenhouse Gases : Gases which trap solar radiation. Of the solar energy entering the earth's atmosphere a portion is reflected back and a portion penetrates onto the earth's surface. The portion reflected back from the earth's surface is at a different wavelength that when it entered. : Carbon dioxide and other gases, which pass solar radiation, absorb this reflected radiation, increasing the earth's temperature. This is much like a greenhouse, hence the name.

Grid Bias : A constant fixed potential applied between the grid and the cathode of a vacuum tube to establish an operating point.

Grid Current : The current that flows in the grid-to-cathode circuit of a vacuum tube.

Grid-Gap Tuning : A method of changing the center frequency of a resonant cavity by physically changing the distance between the cavity grids.

Grid-Leak Bias : A self-bias provided by a high resistance connected across the grid capacitor or between the grid and cathode.

Grommet : An insulator that covers sharp edges of holes through panels and partitions to protect wire insulation from cut-through damage.

Grommet, Connector : An elastomeric seal used on the cable side of a connector body to seal the connector against contamination and to provide stress relief.

Ground : The point in a circuit used as a common reference point for measuring purposes. To connect some point of an electrical circuit or some item of electrical equipment to earth or to the conducting medium used in lieu thereof. A conducting connection, whether intentional or accidental, between a circuit or piece of equipment and the earth, or some body serving as earth; a place of zero electrical potential.

Ground Bounce : Ground bounce refers to the ringing on an output signal when one or more outputs on the same device are being switched from HIGH to LOW. Ground bounce is associated with the inductance and resistance of the ground connection in the integrated circuit.

Ground Clutter : Unwanted echoes, from surrounding land masses, that appear on a radar indicator.

Ground Fault Interrupter : [GFI] A protective device that detects abnormal current flowing to ground and then interrupts the circuit.

Ground Island : A small copper layer contained on one or more layers of a Printed Wiring Board which is only connected to ground. A Ground Island is always much smaller than a ground plane. Ground Islands may be 100% copper fill or hatched to reduce the amount of copper.

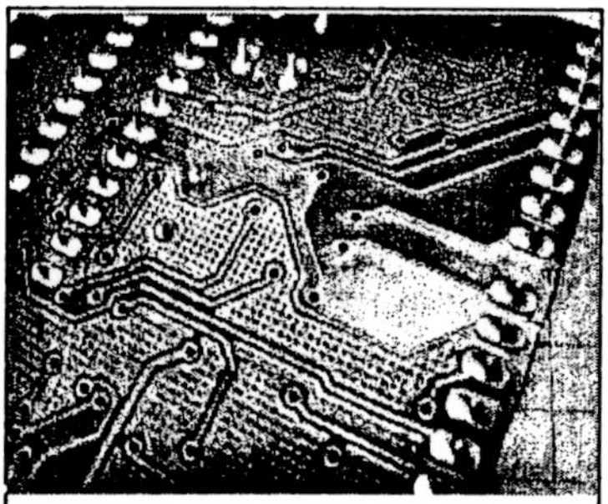

PWB Ground Islands

Ground Pin Current : Ground Pin Current is defined as the amount of current required by the regulator when driving a load. Ground Pin Current varies with regulator input voltage and regulator load. The lower the Ground Pin Current the more efficient the device is. Power consumption of a voltage regulator may be found by multipling the Ground Pin Current and the difference in voltage across the device [Vo/Vi]. [IC General] the current flowing through the ground pin; however this is not a definition in common usage.

Ground Plane : A large copper layer contained on one or more layers of a Printed Wiring Board [PWB] which is only connected to ground. In most cases a ground plane covers most if not all of the PWB layer it resides on. The portion of a ground-plane antenna that acts as ground.

Ground Potential : Zero potential with respect to the ground or earth.

Ground Range : The distance on the surface of the earth between a radar and its target. Equal to slant range only if both radar and target are at the same altitude.

Ground Reflection Loss : The loss of RF energy each time a radio wave is reflected from the earth's surface.

Ground Rod : An earth electrode subsystem installed by the responsible facilities engineering activities at each facility to provide a low resistance path to earth for lightning and power fault currents and ensure that hazardous voltages do not occur within the facility. This subsystem shall be capable of dissipating to earth the energy of direct lightning strokes with no ensuing degradation to itself. This system shall also interconnect all driven electrodes and underground metal objects of the facility.

Ground Screen : A series of conductors buried below the surface of the earth and arranged in a radial pattern. Used to reduce losses in the ground.

Ground Waves : Radio waves which travel near the surface of the earth.

Ground-Controlled Approach : A radar system used to guide aircraft to safe landings in poor visibility conditions.

Ground-Plane Antenna : A type of antenna that uses a ground plane as a simulated ground to produce low-angle radiation.

Groundwater : Water which is contained in geologic strata. : Also properly written as two words, ground water.

Group : A collection of units, assemblies, sub-assemblies, and parts. It is a subdivision of a set or system but is not capable of performing a complete operational function.

Group Velocity : The forward progress velocity of a wave front in a waveguide.

Grown Junction : A method of mixing P-type and N-type impurities into a single crystal while the crystal is being grown.

Guidance Radar : A system which provides information that is used to guide a missile to a target.

Gunmetal : This is a copper-tin alloy, similar to bronze, used where resistance to corrosion and/or wear required. Composition: 88% copper, 8% tin, 4% zinc.

Gyroscope : A mechanical device containing a spinning mass mounted so that it can assume any position in space. The Abbreviation for gyroscope is Gyro.

H

Half-Duplex : Data transmission over a circuit capable of transmitting in either direction, but not simultaneously. Data flow in either direction, but not in both directions at the same time. There are a number of interface buses that only allow Half-Duplex operation, while some bus networks just add a redundant path in the opposite direction to allow communication in either direction at the same time. Operation in both directions at the same time is called Full-Duplex. An Abbreviation for half duplex is HDX.

Half-Flash : An ADC architecture which uses a bank of comparators first to digitise the upper half bits, then uses a digital-to-analog converter (DAC) to subtract that voltage from the input, and then digitises what remains of the input signal to get the lower half bits.

Half-Power Point : A point on a waveform or radar beam that corresponds to half the power of the maximum power point.

Half-Wave Rectifier : A rectifier using only one-half of each cycle to change ac to pulsating dc. Half-Wave Rectifier Circuit.

Half-Wave Voltage Doubler : Two half-wave voltage rectifiers connected to double the input voltage.

Hall Effect : The phenomenon whereby a force is brought to bear on a moving electron or hole by a magnetic field that is applied perpendicular to the direction of motion. The force direction is perpendicular to both the magnetic field and the particle motion directions.

Handover **:** Switching an on-going call to a different channel or cell in a wireless cellular network. Also known as "handoff."

Hand-Shake **:** The interchange of signals between a 'talker' and a 'listener' to exchange data on a bus. During a Handshake a listener slave indicates ready for data, a talker master indicates data ready, the slave then indicates data received.

Hard Disk **:** HD A flat, circular, rigid plate with a magnetisable surface on one or both sides used to store data. Also called a Hard Disk Drive HDD. A hard disk is distinguished from a diskette floppy Disk by virtue of the fact that it is rigid and not considered removable.

Hard Disk

Hardenability **:** A measure of the depth to which a specific ferrous alloy may be hardened by the formation of martensite upon quenching from a temperature above the upper critical temperature.

Hardness **:** The sum of the divalent cation concentrations expressed as meq/L or mg calcium carbonate per liter. It is important because hard waters require increased amounts of soap for bathing or washing clothes and because of scale formation on piping, cooking vessels, water heaters, boilers, heat exchangers, etc.

Hard-Tube Modulator **:** A high-vacuum electron tube modulator that uses a driver for pulse forming.

Harmonic : A frequency that is a whole-number multiple of a smaller base frequency. Of a sinusoidal wave, an integral multiple of the frequency of the wave. Harmonic percentages of a half sine wave; 2nd, 21.2%, 3rd, 0%, 4th 4.2%, 5th, 0% and 6th harmonic is 1.8%.

Harmonic Distortion : The presence of frequencies in the output of a device that are not present in the input signal, and are multiples of components of the input signal. Clipping is a common cause but other nonlinearities can also introduce harmonics.

Harness : An assembly of wires or cables, or wires and cables arranged so it may be installed or removed as a unit in the same electronic or electrical equipment. One or more insulated wires or cables, with or without helical twist; with or without common covering, jacket, or braid; with or without breakouts; assembled with two or more electrical termination devices; and so arranged that as a unit it can be assembled and handled as one assembly.

HB LED : High-Brightness LEDs are any of a new generation of LEDs bright enough for illumination applications such as automotive interior, exterior, and display; room and architectural illumination; task and general lighting; projection display; backlights; and signage.

H-Bridge : A circuit diagram which resembles the letter "H." The load is the horizontal line, connected between two pairs of intersecting lines. It is very common in DC motor-drive applications where switches are used in the "vertical" branches of the "H" to control the direction of current flow, and thus the rotational direction of the motor.

HDLC : High-level Data Link Control is a group of protocols for transmitting [synchronous] data Packets between Point-to-Point nodes. Data is organised into a frame in HDLC. HDLC protocol resides with Layer 2 of the OSI model, the data link layer.

HDSL (High bit-rate Digital Subscriber Line) : The oldest of the DSL technologies, it continues to be used by telephone companies deploying T1 lines at 1.5Mbps and requires two twisted pairs.

HDTV (High-definition Television) : An all-digital system for transmitting a TV signal with far greater resolution than the analog standards (PAL, NTSC, and SECAM). A high-definition television set can display several resolutions, (up to two million pixels versus a common television set's 360,000). HDTV offers other advantages such as greatly improved color encoding and the loss-free reproduction inherent in digital technologies.

Heat Sink : Mechanical device that is thermally-connected to a heat-producing electronic component, designed to conduct heat away from the device. Most heat sinks are aluminum and employ fins to increase surface area and encourage the transfer of heat to the ambient environment.

Heat : If a temperature difference exists between two points heat will flow from the point at the higher temperature to the point at the lower temperature. This flow will continue until equilibrium has been reached.

Hertz (Hz) : Frequency in cycles per second. Named after H.R. Hertz, the German scientist who discovered electrical oscillations. In the case of Alternating Current the frequency is 60 hz in North America and 50 hz in the UK and the rest of the World.

Heterotrophic : A group of organisms which obtain carbon for synthesis from other organic matter or proteins.

Hexagonal Close-Packed (HCP) : A crystal structure found for some metals. The HCP unit cell is of hexagonal geometry and is generated by the stacking of close-packed planes of atoms.

High Speed Steel : A hard steel used for metal cutting tools as it remains hard even at dull red temperatures. It normally contains:. Tungsten 12 to 18%, Chromium up to 5% and Carbon 0.4 to 0.7%.

High-Side : An element connected between the supply and the load. High-side current sensing applications measure current by looking at the voltage drop across a resistor placed between the supply and the load.

Hindered (Zone) Settling : Settling in which particle concentrations are sufficient that particles interfere with the settling of other particles. Particles settle together as a body or structure with the water required to traverse the particle interstices.

Hole (electron) : For semiconductors and insulators, a vacant electron state in the valence band that behaves as a positive charge carrier in an electric field.

Home RF : Trademarked name for Home Radio Frequency, a networking technology which uses antennae and transmitters to provide wireless home networking via transmitted radio signals.

HomePlug : HomePlug (PowerLine) is an industry-standard method for transmitting data via the power lines. It can transmit audio, video, control signals, etc. HomePlug is a trademark of the HomePlug Powerline Alliance; Powerline is the generic term for the method.

Homopolymer : A polymer having a chain structure in which all mer units are of the same type.

Hot Working : Any metal forming operation that is performed above a metal recrystallisation temperature.

Hot-Swap : A power supply line controller which allows circuit boards or other devices to be removed and replaced while the system remains powered up. Hotswap devices typically protect against overvoltage, undervoltage, and inrush current that can cause faults, errors, and hardware damage.

HSDPA (High-Speed Downlink Packet Access) : HSDPA is a 3G radio interface standard in the HSPA family for wireless and cellular handsets or datacards that increase the datarate and improve the traffic handling of existing UMTS standards.

HSPA (High-Speed Packet Access) : HSPA is a collection of radio interface standards for wireless and cellular handsets or datacards that increase the datarate and improve the traffic handling of existing UMTS standards.

HSSI (High-Speed Serial Interface) : HSSI is a short-distance communications standard for data rates from 2Mbps to 52Mbps.

HSUPA (High-Speed Uplink Packet Access) : HSUPA is a 3G radio interface standard in the HSPA family for wireless and cellular handsets or datacards that increase the datarate and improve the traffic handling of existing UMTS standards.

H-Type T-Junction : A waveguide junction in which the junction arm is parallel to the magnetic lines of force in the main waveguide.

Hub : A multi-port device that acts as a switch or repeater connecting one incoming port to another. The Hub may be wired or wireless.

HVAC (Heating, Ventilation, and Air Conditioning) : Industry term for the systems and technology responsible for the heating, ventilation, and air conditioning in buildings. HVAC systems regulate comfort (temperature and humidity), energy efficiency, and air quality.

Hybrid Circuit : A circuit where passive components (resistors, capacitors) are deposited onto a substrate made of glass, ceramic, or other insulating material. Then the active components (diodes, transistors) are attached to the substrate and connected to the passive components on the substrate with a very fine wire.

Hybrid ICs : Two or more integrated circuit types, or one or more integrated circuit types and discrete components on a single substrate.

Hybrid IC PWB

Hybrid Junction : A waveguide junction that combines two or more basic T-junctions.

Hybrid Network : A communication network that uses more than one network topology.

Hybrid Ring : A hybrid-waveguide junction that combines a series of E-type T-junctions in a ring configuration. When properly terminated, energy is transferred from any one branch into any two of the remaining three branches.

Hydraulic Set Cement : A cement that sets through reaction with water.

Hydrocarbon : Any organic compound composed entirely of carbon and hydrogen. Two examples are methane gas and octane.

Hydrocarbon (HC) Analysis. Measurement of the concentration of unburnt hydrocarbon compounds present in the raw exhaust as an indication of combustion efficiency. Expressed as parts per million (p.p.m.).

Hydrogen Bond : A strong secondary interatomic bond which exists between a bound hydrogen atom (its unscreened proton) and the electrons of adjacent atoms.

Hydrometer : An instrument used to measure specific gravity. In batteries hydrometers are used to indicate the state of charge by the specific gravity of the electrolyte.

Hypereutectoid Alloy : For an alloy system displaying a eutectoid, an alloy for which the concentration of solute is greater than the eutectoid composition.

Hypoeutectoid Alloy : For an alloy system displaying a eutectoid, an alloy for which the concentration of solute is less than the eutectoid composition

HyperTransport : A Point-to-Point interface with at least two unidirectional links. The HyperTransport has a data rate of 800M Bytes/ps using 8 bit pairs and a 400MHz clock.

Hysteresis. A characteristic often built in to switching systems so that the switch on and switch off points are separated slightly to prevent oscillation between states. A speed sensitive function such as over-run cut off is an example in which the cut-in speed might be 1200 r.p.m. And the cutout speed 1500 r.p.m.

Hysteresis Loop : [B-H Loop]. A Hysteresis loop showing the relationship between [B] Flex Density and [H] Magnetizing Force.

Hysteresis Loss : The power loss in an iron-core transformer or other alternating-current device as a result of magnetic hysteresis.

HZ : Hertz. A unit of frequency equal to one cycle per second.

I

I/Q : (1) I/Q modulation is a method for combining two channels of information into one signal so that they can be separated at a later stage. Two quadrature carriers, 90 degrees out of phase, are modulated, then combined. Abbreviated from "in-phase/quadrature-phase" which refers to the two carrier signals' phase relationship. (2) IQ (Q should be subscripts but sometimes printed as "IQ" without subscripting).

I2C : Originally designed to be a battery control interface, now used in micro-controller based [uP] professional, consumer and telecommunications control, diagnostic and power management buses. The I2C bus uses a bi-directional Serial Clock Line [SCL] and Serial Data Lines [SDA]. Three speed modes are specified: Standard; 100kbps [Bits per Second], Fast mode; 400kbps, High speed mode 3.4Mbps

I2R Loss : I-Squared-R losses, or Current [squared] x Resistance. Also called Copper Losses. The power lost because of the current through a resistance.

I^2S : Inter-IC Sound (I^2S) is an electrical bus interface standard used for connecting digital audio devices. The I^2S bus separates clock and data signals, resulting in a very low-jitter connection. The bus consists of three lines: a clock line, a word-select line, and a multiplexed-data line.

IBO (Input Back-Off) : In a power amplifier, a measure of how far you must reduce the input power in order to receive the desired output linearity and power. Stated differently, the ratio between the input power that delivers maximum power to the input power that delivers the desired linearity.

IC : (1) Integrated circuit: A semiconductor device that combines multiple transistors and other components and interconnects on a single piece of semiconductor material. (2) Internally Connected.

IC Synchors : Obsolete, synchros with reverse rotation and limited torque capabilities.

IDE (Integrated Drive Electronics) : A series of standards released between 1999 and 2005 which defined the electrical and mechanical interfaces between a personal computer's mother board and a Hard Disk Drive. Maximum bus speed of the IDE bus is 133MBps over an 18 inch Parallel cable. The IDE interface is also known as the ATA interface. The ATA interface was rendered obsolete by the introduction of the Serial ATA [SATA] interface in 2005.

Ideality Factor : A constant adjustment factor used to correct for discrepancies between an ideal PN junction equation and a measured device.

Idempotent Law : In Boolean algebra, combining a quantity with itself either by logical addition or logical multiplication will result in a logical sum or product that is the equivalent of the quantity (for example, A + A = A; A • A = A).

Idle Mode™ : A method for improving the efficiency of switching regulators by skipping pulses when the circuit is lightly loaded. This variation in PWM (pulse-width modulation) combines the efficiency at low loads afforded by PFM (Pulse-Frequency Modulation) with PWM's efficiency and low-noise characteristics at higher loads. At light loads the circuit skips pulses as necessary (acting like a PFM circuit). At higher loads it acts like PWM. The net result is the maximum efficiency over the widest possible load range.

Idle Speed Control System : An electronically controlled system which compensates for changes of temperature, load etc. to maintain a predetermined idle speed via an electrically operated valve - often using a stepper motor. May also alter ignition timing to obtain faster response to changing loads.

Idle-Channel Noise : Noise that is present in a communications channel when no signals are applied. The channel conditions and terminations must be stated for idle-channel noise measurements to be meaningful.

Idler Frequency **:** In a parametric amplifier, the difference between the input signal and the pump signal frequency.

IEC **:** (1) IEC stands for the International Electrotechnical Commission: An "organisation that prepares and publishes international standards for all electrical, electronic and related technologies." (2) Commonly used to refer to one of the 13 power connectors described by specification IEC 60320. Most commonly refers to the C13 and C14 connectors used by most computers and many AC-powered electronic devices to connect the AC power. (3) Integrated electronic component.

IEEE **:** The IEEE (Eye-triple-E) is a non-profit, technical professional association of more than 360,000 individual members in approximately 175 countries. The full name is the Institute of Electrical and Electronics Engineers, Inc., although the organisation is most popularly known and referred to by the letters I-E-E-E. IEEE also sponsors many electrical and electronic standards.

IEEE-1284 **:** A personal Computer interface that defined a parallel printer bus specification at 1MBps. IEEE-1284 defines a Point-To-Point asynchronous bi-directional interface. Devices may be either 1284 compatible {the older parallel port devices} or 1284 compliant. The maximum length for a printer cable is 25 feet, 32 feet maximum. Centronics parallel cables run out to 12 feet. The IEEE1284 cable replaced the 'Centronics' cable which is obsolete.

IF (Intermediate Frequency) **:** Radio communications systems modulate a carrier frequency with a base band signal in order to achieve radio transmission. In many cases, the carrier is not modulated directly. Instead, a lower IF signal is modulated and processed. At a later circuit stage, the IF signal is converted up to the transmission frequency band.

IF Amplifier **:** Usually a narrow-bandwidth IF amplifier that is tuned to one of the output frequencies produced by the mixer.

Ignition Map : A three dimensional representation of the relationship between ignition advance and sensor inputs, normally considered in terms of engine speed and load.

IIP3 (Third Order Input Intercept Point) : The point at which the power in the third-order product and the fundamental tone intersect, when the amplifier is assumed to be linear. IIP3 is a very useful parameter to predict low-level inter modulation effects.

IMA (Inverse Multiplexing over ATM) : An MGX card module that supports T3 or E3 inverse multiplexing on up to eight T1 or E1 lines.

Image Frequency : Receivers typically convert RF signals to a lower Intermediate Frequency (IF) for demodulation. In addition to the IF, a second signal, called the "image frequency" is often generated and filtered out.

Image Rejection : The measure of a receiver's ability to reject signals at its image frequency. It is normally expressed as the ratio, in dB, of the receiver's sensitivity at the desired frequency versus the sensitivity at the image frequency.

IMD (Intermodulation Distortion) : When two signals mix in non-linear circuits or devices, new frequency components are created that are not in the original signal. The resulting signal error is called intermodulation distortion, or IMD.

Impact Energy : A measure of the energy absorbed during the fracture of a specimen of standard dimensions and geometry when subjected to very rapid (impact) loading. Charpy and Izod impact tests are used to measure this parameter, which is important in assessing the ductile-to-brittle transition behavior of a material.

Impedance : Impedance, represented by the symbol Z, is a measure of the opposition to electrical flow. It is measured in ohms. For DC systems, impedance and resistance are the same, defined as the voltage across an element divided by the current ($R = V/I$). In AC systems, the "reactance" enters the equation due to the frequency-dependent contributions of capacitance and inductance. Impedance in an AC system is still measured in ohms and represented by the equation Z = V/I, but V and I are frequency-dependent.

Imperial Standard Wire Gauge : Tabulated wire diameters presented in imperial and metric.

Implosion : The inward bursting of a CRT because of high vacuum. The opposite of explosion.

Impulse Waveform : A short duration pulse. The pulse duration is defined from the peak duration to 50 percent of peak.

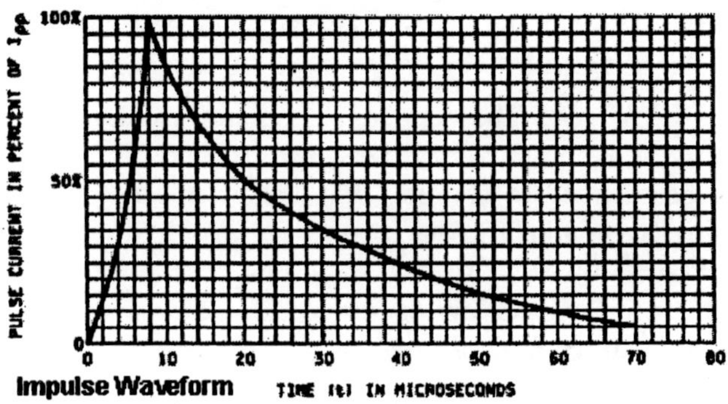

Impulse Duration

IMVP (Intel Mobile Voltage Positioning) : A technology in which the processor voltage (VCC) is dynamically adjusted, based on the processor activity, to reduce processor power. It allows higher processor clock speed at a given power consumption; or lower consumption at a given clock frequency.

In Situ Treatment : Treatment of a waste in place, as opposed to pumping or digging the waste up and then treating it.

Incident Wave : The wave that strikes the surface of a medium. The wave that travels from the sending end to the receiving end of a transmission line.

In-Circuit Meter : A meter permanently installed in a circuit; used to monitor circuit operation.

In-Phase : Applied to the condition that exists when two waves of the same frequency pass through their maximum and minimum values of like polarity at the same instant. Voltage and current are in-phase across a resistor but out of phase in a capacitor or Inductor. Note the maximum value of the current wave or voltage may be any value, as long as they reach their maximum point at the same time.

Incoherent : Refers to radiation on a broad band of frequencies.

Inductance : A property of electricity whereby the magnetic field created by an electric current flowing through a coil of wire opposes any change in that current. The effect is of making an electric current behave as if it had inertia.

Inductive Kickback : The very rapid change in voltage across an inductor when current flow is interrupted. Snubber diodes are often used to channel this energy in relays, and other inductive loads. Kickback can be a problem (causing EMI and component failure); or it can be used in power supply circuits to develop higher or opposite-polarity voltages from a single supply.

Inertia Switch : A shock sensing switch which is commonly used to cut off power to the injection system in the event of the vehicle suffering an impact. Must be reset before the vehicle can be used again.

Infiltration : The movement of water from the surface of the land through the unsaturated zone and into the groundwater. This occurs during and immediately after precipitation events. : It can also occur at the bottom of lakes and rivers.

InfiniBand : InfiniBand architecture is an industry standard, channel-based, switched-fabric, interconnect architecture for servers. InfiniBand architecture changes the way servers are built, deployed, and managed.

Influent : The fluid entering a system, process, tank, etc.An effluent from one process can be an influent to another process.

Ingress Protection : An Ingress Protection (IP) rating indicates how well an enclosure is protected from penetration by contaminants such as dust or fluids (such as water).

Injector (Electric Type) : Precision solenoid operated valve controlling flow of fuel admitted to engine inlet ports according to pulses from the ECU.

Inline Engine Configuration : This is a model two-stroke engine with a displacement of 0.5 cu. in. (built by N. Taylor)

Input : The current, voltage, power, or driving force applied to a circuit or device. The input signal applied to an IC.

Input CMVR (V) : Common-mode voltage range (CMVR) or Input Voltage Range (IVR): For signal processing devices with differential inputs, such as an op amp, CMVR is the range of common mode signal for which the amplifier's operation remains linear. If we let the voltage present on the "-" input equal V1, and the voltage on the "+" input equal V2, then the common mode voltage is VCM = (V1+V2)/2. Some op amps, for instance, will only allow the common mode voltage of a signal to come within a diode drop or so of the power supply rails.

Input End : The end of a two-wire transmission line that is connected to a source.

Input Impedance : Impedance presented to the transmitter by the transmission line and its load.

Input/Output : Pertaining to either input or output or both, especially in data processors.

Inrush Current : A momentary input current surge, measured during the initial turn-on of the power supply. This current reduces to a lower steady-state current once the input capacitors charge. Hotswap controllers or other forms of protection are often used to limit inrush current, because uncontrolled inrush can damage components, lower the available supply voltage to other circuits, and cause system errors.

Insertion Loss : The loss in power caused by the insertion of a network between a source and a load.

Instantaneous Amplitude : The amplitude at any given point along a sine wave at a specific instant in time.

Instantaneous Automatic Gain Control (IAGC) : A circuit that can vary the gain of the radar receiver with each input pulse to maintain a nearly constant output peak amplitude.

Instantaneous Value : The magnitude at any particular instant when a value is continually varying with respect to time.

Instrinsic Semiconductor : A semiconductor material for which the electrical behavior is characteristic of the pure material.

Instrumentation Amplifier : A differential amplifier utilizing an input buffer on both the positive and negative input lines. This example uses operational amplifies and its associated resistors required to produce gain.

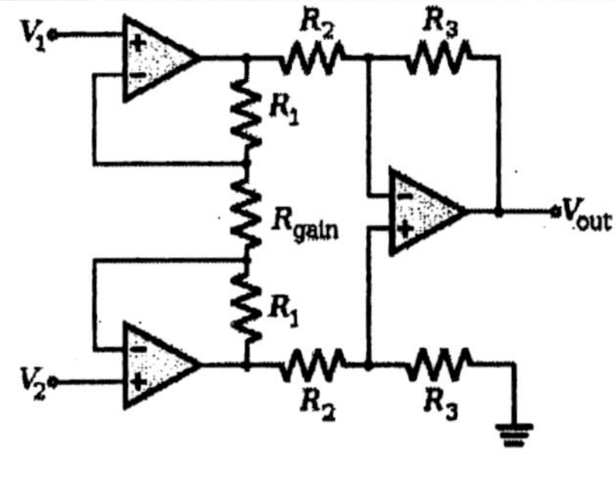

Instrumentation Amplifier

Insulation : A material used to prevent the leakage of electricity from a conductor and to provide mechanical spacing or support as protection against accidental contact with the conductor.

Insulation Displacement Connection (IDC) : A connector that cuts through the insulation of a conductor to make contact with the wire.

Insulation Resistance : The resistance offered by an insulating material to current leakage.

Insulator (electrical) : A nonmetallic material that has filled valence band at 0 K and a relatively wide energy band gap.

Insulator : Material of such low conductivity that the flow of current through it can usually be neglected. A device having high electrical resistance; used for supporting or separating conductors so as to prevent undesired flow of current from the conductors to other objects.

Integral Nonlinearity : A measure of a data converter's ability to adhere to an ideal slope in its transfer function. It can be specified using end-point or best-straight-line fit.

Integrated Circuit (IC) : A circuit in which many elements are fabricated and interconnected by a single process (into a single chip), as opposed to a "nonintegrated" circuit in which the transistors, diodes, resistors, and other components are fabricated separately and then assembled.

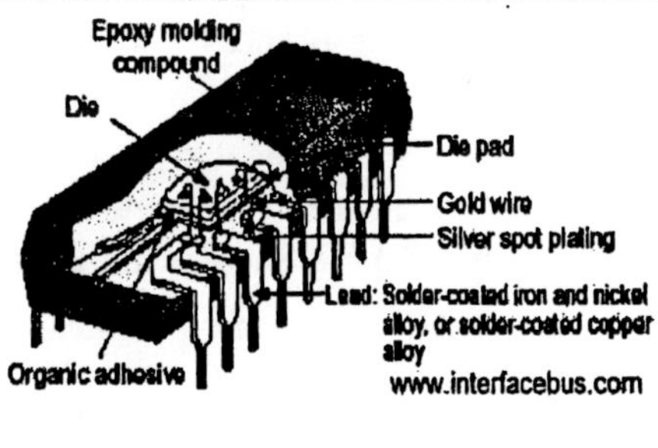

Integrated Circuit

Integrated Heat Spreader : An Integrated Heat Spreader (IHS) is the surface used to make contact between a heatsink or other thermal solution and a CPU or GPU processor.

Integrator : A circuit that takes performs an integration on the input signal. The circuit example shows an Operational Amplifier used as an integrator.

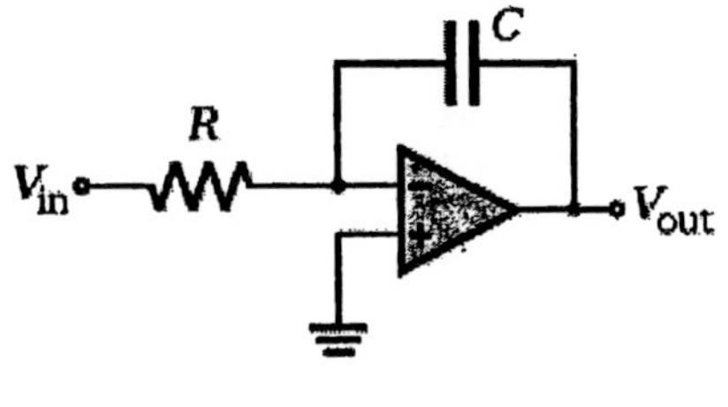

Integrator

Intellectual Property : Creations of the intellect such as trade knowledge, technical information, and literary or artistic work, including patents, copyrights, and trademarks.

Interleave : To organise the data sectors on a computer hard disk, so the read/write heads can access information faster.

Intermetallic : A compound of two metals that has a distinct chemical formula. The bonds in intermetallic compounds are often partly ionic.

Intermodulation : A process whereby signals mix together in a circuit and nonlinearities in the circuit create undesired output frequencies that are not present at the input.

Internet Protocol : Standard method for data transfer used on the Internet.

Invar : This is a trade name for iron nickel alloy. 36% nickel, 63.8% iron and 0.2% carbon. Used where small coefficient of expansion is required e.g. measuring tapes, pendulums.

Invariant Point : A point on a binary phase diagram at which three phases are in equilibrium.

Inverter : A circuit with one input and one output. Its function is to invert or reverse the input. When the input is high, the output is low, and vice versa. The inverter is sometimes called a NOT circuit, since it produces the reverse of the input.

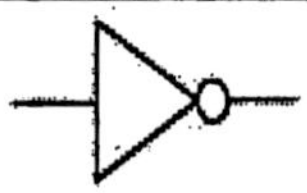

Inverter

Inverting Switching Regulator : A switch-mode voltage regulator in which output voltage is negative with respect to its input voltage.

IO-Link : IO-Link is a 24-volt, three-wire, half-duplex, point-to-point sensor and actuator communication interface. Remote configuration, diagnostics, event triggering and process data readout are made possible from a PLC via a three layer protocol stack. IO-Link can be used for simple binary sensors and smart sensors.

Ion Exchange : An adsorption process in which one ion is exchanged for another ion of like charge. There is an equivalence of exchanged charge.

IR (Infrared) : Light that has a frequency below the visible light spectrum, used for remote controls, line-of-sight wireless data, and night vision applications, among others.

IrDA (Infrared Data Association) : A group of device manufacturers that developed a standard for transmitting data via infrared light waves.

IRE (Institute of Radio Engineers) : IREs are units of measurement dividing the area from the bottom of sync to peak white level into 140 equal units. 140 IRE = 1VP-P

Irreversible Reaction : A reaction in which the reactant(s) proceed to product(s), but the products react at an appreciable rate to reform reactant(s).

ISI (Inter-Symbol Interference) : A form of interference that occurs when echoes of a radio-signal interfere with the original signal. ISI can reduce the effective data rate of wireless LAN transceivers.

ISM (Industrial, Scientific and Medical) : Radio frequency bands made available for use by communication equipment without license, within certain maximum emitted power limits. Equipment which uses the ISM band must tolerate interference from other such equipment. Common uses include WiFi (802.11a, b, and g) and cordless phones.

Isomers : Two or more different compounds with the same chemical formula but different structure and characteristics.

Isotactic : A type of polymer chain configuration wherein all side groups are positioned on the same side of the chain molecule.

Isotope : Atoms of the same element having the different masses.

Izod Impact Test : One of two tests that may be used to measure the impact energy of standard notched specimen.

J

Jabber : A condition wherein a station transmits for a period of time longer than the maximum permissible packet length, usually due to a fault condition.

Jack : A socket. A receptacle.

Jacket : The outermost layer of insulating material of a cable or harness. The outer sheath which protects a cable. The outer sheath or covering material, such as natural or synthetic rubber, lead sheath,steel tape,or a thermoplasitc compound, applied over a single-insulated conductor or over an assembly of insulated conductors for protection against crushing,cutting,and abrasion of the conductors and their insulation.

Jacketed Cable : A bundle of insulated wires encased in a common sheath.

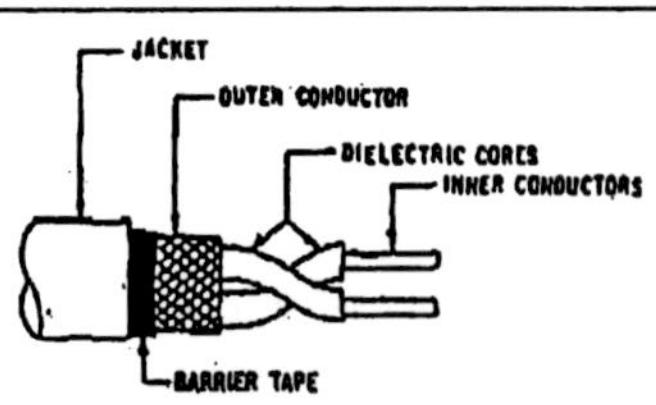

Jacketed Twisted Pair Cable

JAN : Joint Army Navy specification.

JBOD (Just a Bunch of Disks) : An array of hard disks without a controller.

JEDEC : Joint Electron Device Engineering Council

JFET : A JFET, or junction field-effect transistor, or JUGFET, is a FET in which the gate is created by reverse-biased junction (as opposed to the MOSFET which creates a junction via a field generated by conductive gate, separated from the gate region by a thin insulator). For example: A p-channel JFET would consist of a bar of p-type silicon with the "drain" at one end and the "source" at the other. Between these two terminals is some n-type material connected.

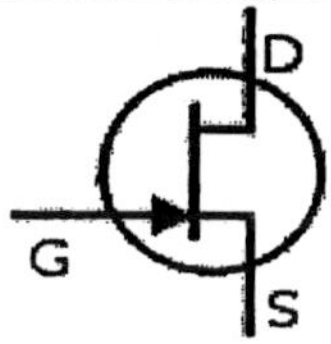

JFET

Jitter : The slight movement of a transmission signal in time or phase that can introduce errors and loss of synchronisation. More jitter will be encountered with longer cables, cables with higher attenuation, and signals at higher data rates. Also, called phase jitter, timing distortion, or intersymbol interference.

JK Flip Flop : A style of flip flop that has a input and a K input, and also may have a Set [S] and Reset [R] input. No change occurs when J and K are low. When the K is low the output is low and when the J is low the output goes high. When both J and K are high the output toggles.

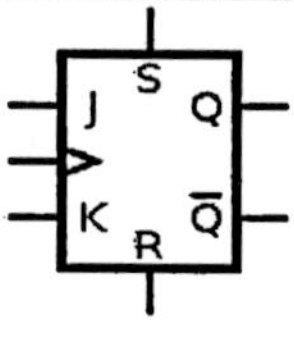

JKFFF

Joule : A measurement of energy or work. In mechanical systems, it's the a force of one newton, moving an object a distance of one meter. In electronics, it's the same amount of energy, in electrical units. One joule is one watt of power, applied for one second (a watt-second); or a coulomb of electrical charge raised to a potential of one volt.

Joystick : A peripheral device used with personal computers to translate physical movement in two axis into electrical signals used by the computer.

JPEG : Joint Photography Experts Group; more commonly, files that are compressed using the JPEG standard.

JTAG : Joint Test Action Group. IEEE Std 1149.1-1990, Test Access Port and Boundary-Scan Architecture (3) The JTAG interface is a serial bus with four signals: Test Clock (TCK), Test Mode Select (TMS), Test Data Input (TDI), and Test Data Output (TDO).

Jumper : A short length of wire used to complete a circuit. A small plastic case with two female receptacles [shunt] used to short two adjacent contacts or pins, normally on 0.1 inch centers. The term may also describe a Zero Ohm Resistor. A jumper is usually used with a 2-pin header.

Junction Box : A box with a cover that serves the purpose of joining different runs of wire or cable and provides space for the connection and branching of the enclosed conductors.

Junction Diode Sensor : The use of a PN junction on a silicon die for determining die temperature.

Junction Diode : A two-terminal device containing a single crystal of semi-conducting material that ranges from P-type at one terminal to N-type at the other.

Junction Transistor : A bipolar transistor constructed from interacting PN junctions. The term is used to distinguish junction transistors from other types, such as field-effect and point-contact.

Junction : The connection between two or more conductors. The contact between two dissimilar metals or materials, as in a thermocouple.

K

k **:** 1. Kilo: Metric unit representing 1000. E.g.: 1kHz is a 1 kilohertz (1000 Hertz). Note that the k is always lowercase. In digital systems, "K" or "k" is often used to mean 210, that is, 1024. This is not well-standardised but it's usually apparent from context.

Kanal + **:** Kanal+ support allows a VCR to record audio and video signals captured by both the set top box (STB) and the television, without changing SCART connections on the back of the TV, STB, and VCR.

Karnaugh Map **:** A graphical method of representing a Boolean function. A True Table displayed so that a circuit reduction can take place.

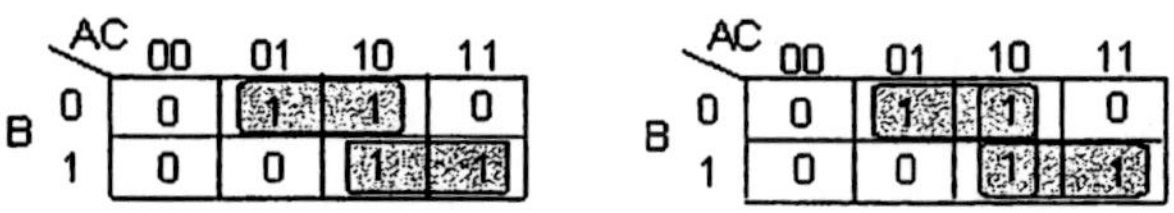

Sample Karnaugh Map

Keep-Alive Voltage **:** DC voltage applied to a tr gap electrode to produce a glow discharge that allows the tube to ionise faster when the transmitter fires.

Keep-Out Zone **:** The area on or near a CPU or GPU processor that the circuit board layout design can not use, due to thermal management components, cooling, and mounting constraints.

Kelvin : Temperature scale. Zero K is defined as absolute zero. 273.15K is 0 degrees C. Note that temperatures on the kelvin scale are called kelvins, not "degrees kelvin." The K symbol is uppercase and used without a degree symbol. The word "kelvin" in this context is not capitalized.

Kernel : A module of a program that forms a logical entity or performs a unit function.

Kerogen : A fossilised organic material present in oil shale and some other sedimentary rocks.

Ketones : Organic compounds with two hydrocarbon groups bonded to a carbonyl group.

Key Clicks : Interference in the form of "clicks" or "thumps" caused by the sudden application or removal of power.

Keyboard : A peripheral device used with a personal computer which allows data entry.

Key-Click Filters : Filters used in keying systems to prevent key-click interference.

Keyed-Oscillator Transmitter : A transmitter in which one stage is used to produce the RF pulse.

Keyer : A device that changes dc pulses to mark and space modulation for teletypewriter transmissions. A synchronizer. This term is obsolete.

Keying Relays : Relays used in radio transmitters where the ordinary hand key cannot accommodate the plate current without excessive arcing.

Kinetic Energy : Energy that a body possesses by virtue of its motion.

Kirchhoff's Laws : The algebraic sum of the current flowing toward any point in a circuit and the current flowing away from it is zero. The algebraic sum of the products of the current and resistance in each of the conductors in any closed path in a network is equal to the algebraic sum of the electromotive forces in the path.

Klystron Power Amplifier **:** A multicavity microwave electron tube that uses velocity modulation.

Knee of the Curve **:** The point of maximum curvature of a magnetisation curve.

Knock Sensor **:** A type of microphone or accelerometer which, by detecting the shock waves created when an engine suffers detonation, activates a control circuit which causes ignition to retard slightly until the condition ceases.

kVM (Keyboard Video Mouse) **:** Defacto standard for the three cables used on a typical cpr: One for the keyboard; one for the monitor (video); one for the mouse. Also: A KVM switch is a switch box used to connect one KVM to multiple computers.

kW **:** Kilowatt (or kilowatts): 1000 watts.

L

Laf Sensor : A variant of the Lambda sensor developed by Honda which makes it possible to measure air/fuel ratio over a wide range, thereby making possible a Lambda feedback lean-burn system.

Lambda Feedback : A mode of operation in which the signal from a Lambda sensor is used to control fuelling to a predetermined air/fuel ratio. Mainly used with three-way catalyst systems requiring precise control of stoichiometric fuelling.

Lambda Sensor : A device located in the exhaust stream which produces a sharp change of voltage according to the presence or absence of oxygen, enables the ECU to determine that the mixture is weaker or richer than stoichiometric. Generally used in conjunction with a Three Way Catalyst.

Lambda : Greek symbol used to denote air/fuel ratio on basis that for no air L = 0, for stoichiometric L = 1, etc. As an example L = 1.1 would be a fairly typical weak cruise mixture.

Laminate : A product made by bonding two or more layers together, usually of different materials, under heat and pressure to form a single structure.

Laminated Core : A core built up from thin sheets of metal insulated from each other and used in transformers.

LAN (Local Area Network) : A computer network, usually within one building, that connects computers, file and mail servers, storage, peripherals, and other devices in a way that permits data interchange and sharing of resources. Ethernet and WiFi (802.11) are common examples.

Landfill : A legal and controlled area for the placement of wastes into the ground.

Landfilling : The placement of wastes into the land under controlled conditions to minimise their migration or effect on the surrounding environment.

Lands : Conductors or runs on Printed Circuit Boards [PCB's]. However runs are normally called Traces, while pads are referred to as lands.

Lap Splice : A joint formed by the connecting of two or more conductors. The Lap splice is one of a number of different styles of wire splices.

Wire Lap Splice

Lap Winding : An armature winding in which opposite ends of each coil are connected to adjoining segments of the commutator so that the windings overlap.

Larch : An extremely durable wood used for rough work. It warps a lot when dried, but can be polished to a fine finish when dry.

Laser Driver : An IC that supplies modulated current to a laser diode in response to an input serial-data stream.

Laser : An acronym for light amplification by stimulated emission of radiation.

Latch : A circuit designed to store information. The output of a latch will change when the clock goes high but will not latch until the clock goes low. As a normal process in the design a Latch will not be permitted to passed a design review. Normally when found in an FPGA a latch will not pass a design review, Latches are considered bad design practice because they are unusable in synchronous logic circuits.

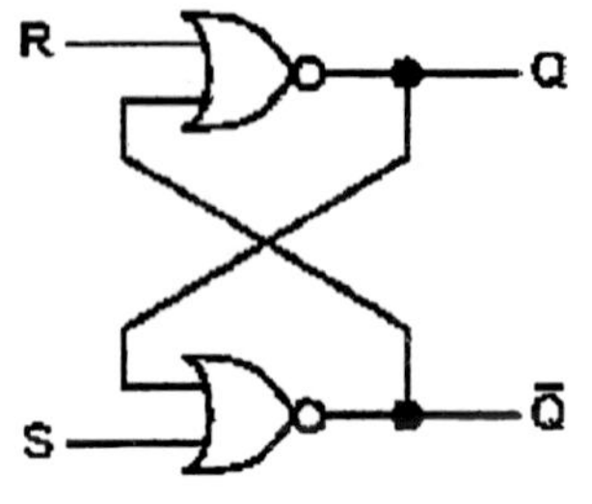

NOR Latch

Lathe Tools : The cutting and clearance angles of a lathe tool will be dependent on the type of material they are to be used on and the required quality of finish.

Lathes : Lathes are used for many engineering operations: boring, facing, screwcutting.

Lattice Parameter : The combination of unit cell edge lengths and interaxial angles that defines the unit cell geometry.

Lattice : The regular geometrical arrangement of points in crystal space.

L-Band : The group of radio frequencies extending from 390MHz to 1550MHz. The GPS carrier frequencies (1227.6MHz and 1575.42MHz) are in the L-band.

LC Capacitor-Input Filter : This is the most common type of filter. It is used in a power supply where output current is low and load current is relatively constant.

LC Choke-Input Filter : This filter is used in power supplies where voltage regulation is important and where the output current is relatively high and subject to varying load conditions.

LCC : (1). Leadless Ceramic Chip Carrier or Leadless Chip Carrier: An IC package, usually ceramic, that has no leads (pins). It instead uses metal pads at its outer edge to make contact with the printed circuit board. (2). Leaded Chip Carrier, also called PLCC or Plastic Leaded Chip Carrier: A square surface mount chip package in plastic with leads (pins) on all four sides.

LDO (Low Drop Out) : A linear voltage regulator that will operate even when the input voltage barely exceeds the desired output voltage.

Leachate : A liquid generated in landfills.It is the result of water seeping into and through the wastes.As the water contacts the waste materials it dissolves part of the organic and inorganic matter contained in the landfill.If this leachate is allowed to exit the bottom of the landfill, it will carry contaminants to the groundwater and/or adjoining surface water.

Leaching : The act of dissolving the soluble portion of a solid mixture by some solvent. : An example is the dissolving of inorganic or organic contaminants from refuse in a landfill by infiltrating rain water.

Lead Inductance : The inductance of the I/O pins of an IC. The inductance of the lead wires connecting the internal components of an electron tube. Lead Inductance in IC pins can cause Ground Bounce.

Lead Sheath : A continuous jacket of lead molded around a single conductor or multiple conductor cable. Generally used to ensure conductors are protected from water or extensive moisture.

Lead-Acid Cell : A cell in an ordinary storage battery in which electrodes are grids of lead containing an active material consisting of certain lead oxides that change in composition during charging and discharging. The electrodes or plates are immersed in an electrolyte of diluted sulfuric acid.

Leakage Current : The small amount of current that flows through the dielectric between the conductors of a transmission line. Leakage current is that current which flows through the equipment conductive paths to a solidly grounded source.

Leakage Flux : Magnetic flux lines produced by the primary winding that do not link the turns of the secondary winding.

Leakage Inductance : Leakage inductance in a transformer is an inductive component that results from the imperfect magnetic linking of one winding to another. In an ideal transformer, 100% of the energy is magnetically coupled from the primary to the secondary windings. Imperfect coupling reduces the signal induced in the secondary windings. The electrical equivalent is some self-inductance in series with the primary windings that are properly coupled. This series inductance is the "leakage inductance."

Leakage Resistance : The electrical resistance that opposes the flow of current through the dielectric of a capacitor. The higher the leakage resistance, the slower the capacitor discharges or leaks across the dielectric.

Lean Burn Combustion : A combustion system which enables an engine to burn very weak mixtures in pursuit of good fuel efficiency and low exhaust emission levels. Usually involves good mixture preparation and carefully controlled swirl and turbulence.

Least Significant Digit (LSD) : The LSD is the digit whose position within a given number expression has the least weighting power.

LED (Light-Emitting Diode) : A semiconductor device that emits light (usually visible or infrared) when forward-biased. The application note, "Driving LEDs in Battery-Operated Applications: Controlling Brightness Power Efficiently" has a good explanation of how LEDs work, especially with regard to current vs. LED brightness and schemes for matching brightness when driving multiple LEDs.

Types of LEDs

Left-Hand Rule for Generators : A rule or procedure used to determine the direction of current flow in a generator.

LEG : A current-carrying conductor intended to deliver power to or from a load normally at an electrical potential other than ground.

Legacy : An out-of-date system or component that is still being used. Which could also be a device or component still being used after it's useful life.

Lenz's Law : The current induced in a circuit, caused by its motion in a magnetic field or a change in it's magnetic flux, in such a direction as to exert a mechanical force opposing the motion or to oppose the change in flux.

Letter Drills : This system of numbering drills is no longer used, but may be of use in deciphering old drawings.

Level Translator : A device which translates a logic signal from one type to another, for example, ECL to TTL.

Lever Rule : Mathematical expression whereby the relative phase amounts in a two-phase alloy at equilibrium my be computed.

LFSR (*Linear Feedback Shift Register*) : A shift register in which some of its outputs are connected to the input through some logic gates (typically, an exclusive-or (XOR). A wide variety of bit patterns can be generated inexpensively, including pseudo-random sequences. Can be used as a noise generator.

Life Cycle : The stages that a product goes through from conception to obsolescence, following a bell curve. There are six distinct phases of a Products Life Cycle: Stage 1; Product Introduction; Stage 2; Product Growth [rapid sales growth]; Stage 3; Product Maturity [sales grow stable, left of curve peak]; Stage 4. Product Saturation [sales begin to level out, right of peak]; Stage 5. Product Decline [Sales begin to decline, Obsolete]; Stage 6. Product Phase-Out Sales continue decline, Lifetime buys]. Products in either stage 5 or 6 are not recommended for new designs. Stage 6 is also called End-Of-Life.

Ligand **:** The ion or molecule which surrounds or complexes with the central atom or ion.

Light Rays **:** Light waves emitting from a source in straight lines.

Light-Emitting Diode (LED) **:** A PN-junction diode that emits visible light when it is forward biased. Depending on the material used to make the diode, the light may be red, green, or amber

Lighthouse Tube **:** An electron tube shaped like a lighthouse that is designed to handle large amounts of power at uhf frequencies.

Light-Pipe **:** A cylindrical tube or fiber used to pass light from one location to another.

Limp Home Mode **:** The capability of some advanced systems to revert to a secondary simplified control of fuelling or ignition in the event of a fault developing in either the ECU or its sensors.

LIN (Local Interconnect Network) **:** Defined by the LIN-BUS consortium, a LIN is a low data-rate, single-wire communications system, used in automotive and heavy vehicle applications.

Lineman splice **:** A joint formed by the connecting of two or more conductors. The Lineman splice is one of a number of different styles of wire splices.

Lineman splice

Line Regulation **:** The ability of a power-supply voltage regulator to maintain its output voltage despite variations in its input voltage.

Linear Mode **:** Uses a linear-pass element (BJT or FET) to control/ regulate the charging voltage/current.

Linear Regulator : A voltage regulator that is placed between a supply and the load and provides a constant voltage by varying its effective resistance.

Lithium Batteries : Lithium batteries for low-power, high-reliability, long-life applications such as non-volatile memory and timekeeping (typically in coin-shaped cells) use a variety of lithium-based chemistries (as differentiated from lithium-ion).

Lithium-ion Batteries : A number of battery chemistries are based on the element lithium, a highly-reactive metallic element. Lithium-based batteries are common in two applications: Power for portable equipment such as cell phones, laptops, and MP3 players; and low-power, long-life applications such as powering memory elements and clocks. *Lithium-ion (Li+, Li-Ion, Lion)* cells are generally used as power sources for portable equipment.

Live Centre : A centre that revolves with the work, for example the ball bearing type tailstock centre.

LMDS (Local Multipoint Distribution Service) : A broadband radio service, located in the 28GHz and 31GHz bands, designed to provide two-way transmission of voice, high-speed data and video.

Load Law : The way in which the fuelling or ignition advance changes according to engine load, which can be derived from throttle angle, manifold pressure or mass airflow.

Load Regulation : Load regulation refers to circuitry that compensates for changes in load.

Local Temperature : The temperature measured on the die of the temperature-measuring integrated circuit.

Local Temperature Sensor : An element or function of an integrated circuit that measures its own die temperature.

Logic : The basic principles and applications of truth tables, interconnections of off-on circuit elements, and other factors involved in mathematical computation in automatic data processing systems and other devices.

Logic Circuit : The primary control information processor in digital equipment; made up of electronic gates and so named because their operation is described by simple equations of a specialized logic algebra.

Logic Diagram : In computers and data processing equipment, a diagram representing the logical elements and their interconnections without necessarily expressing construction or engineering details.

Logic Element : The smallest building blocks that can be represented by operators in an appropriate system of symbolic logic. Typical logic elements are the AND-gate and the flip-flop, which can be represented as operators in a suitable symbolic logic. Also a device that performs the logic function.

Logic Instruction : Any instruction that executes a logic operation that is defined in symbolic logic, such as AND, OR, NAND, or NOR.

Logic Operation : A non-arithmetical operation in a computer, such as comparing, selecting, making references, matching, sorting, and merging, where the logical YES or NO quantities are involved.

Logic Switch : A diode matrix or other switching arrangement that is capable of directing an input signal to one of several outputs.

Logic Symbol : A symbol used to represent a logic element graphically. Also a symbol used to represent a logic operator.

Long Haul : A network that spans distances larger than a local area network (LAN). Because electrical and optical transmissions fade over distance, long-haul networks are difficult and expensive to implement.

Long Term Evolution : Proposed next-generation, wireless, broadband network. LTE is presently a project within 3GPP, the Third Generation Partnership Project.

Longitudinal Mode : A propagation mode in which the two conductor of a pair act as a single conductor.

Longitudinal Waves : Those waves in which the disturbance (back and forth motion) takes place in the direction of propagation. Sometimes called compression waves.

Loop : A curved conductor that connects the ends of a coaxial cable or other transmission line and projects into a waveguide or resonant cavity for the purpose of injecting or extracting energy.

Loose Coupling : Inefficient coupling of energy from one circuit to another that is desirable in some applications. Also called weak coupling.

Low Frequency : The band of frequencies from 30 kHz to 300 kHz.

Low Power Fm Radio : [LPFM] A broadcast service that permits the licensing of 50-100 watt FM radio stations within a service radius of up to 3.5 miles and 1-10 watt FM radio stations within a service radius of 1 to 2 miles.

Lowest Replaceable Unit : An LRU is the lowest possible unit to be replaced within the system component during site level maintenance activities. It is a separate, installable physical package performing a single function or group of closely related functions.

Lowest Usable Frequency : The minimum operating frequency that can be used for communications between two points.

Low-Level Modulation : Modulation produced in an earlier stage than the final.

Low-Pass Filter : A filter that passes a majority of the low frequencies on to the next circuit and rejects, or attenuates, the higher frequencies. Also called a high-frequency discriminator.

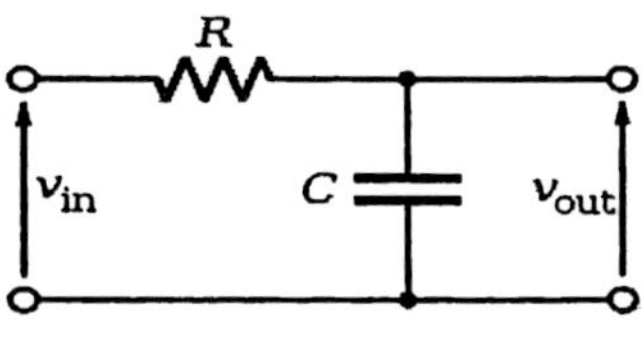

Low-Pass Filter

Low-Side : An element connected between the load and ground. Low-side current sensing applications measure current by looking at the voltage drop across a resistor placed between the load and ground.

LSB (Least-significant bit) : In a binary number, the LSB is the least weighted bit in the number. Typically, binary numbers are written with the MSB in the left-most position; the LSB is the furthest-right bit.

Luminance : (1) The emitted light, projected per unit area, measured in cd/m^2 (candela per square meter). Often incorrectly equated with "brightness". (2). The black and white portion of a video signal, also referred to as the "Y" component. A composite, Y/C, or Y/Pb/Pr video signal combines a luminance signal with color components.

Lumped Constants : The properties of inductance, capacitance, and resistance in a transmission line.

Lumped Impedance Tuning : The insertion of an inductor or capacitor in series with an antenna to electrically lengthen or shorten the antenna.

M

M (mega) **:** The prefix M or mega indicates the unit should be multiplied by 10^6.

m (milli) **:** The prefix m or milli indicates the unit should be multiplied by 10^{-3}.

M2M **:** Machine-to-machine communications, via wireless technologies such as cell phone network technologies, WLAN, Bluetooth, and RFID (radio frequency identification). Applications include automatic meter reading, fleet management, vending, monitoring and control, security and alarms, and telemedicine.

mA **:** Milliampere, or milliamp: 1/1000 of an Ampere. Ampere is the basic unit for measuring electrical current.

MAC Address **:** Media Access Control Address (maca, MAC): A hardware address that uniquely identifies each node of a network, as in IEEE-802 (Ethernet) networks. The MAC layer interfaces directly with the network medium.

Macromolecule **:** A huge molecule made up of thousands of atoms.

Magnetic Field Strength **:** The intensity of an externally applied magnetic field.

Magnetic Field **:** The region in which the magnetic forces created by a permanent magnet or by a current-carrying conductor or coil can be detected. The field that is produced when current flows through a conductor or antenna.

Magnetic Flux Density **:** The magnetic field produced in a substance by an external magnetic field.

Magnetic Induction : Generating a voltage in a circuit by the creation of relative motion between a magnetic field and the circuit. The relative motion can be the result of physical movement or the rise and fall of a magnetic field created by a changing current.

Magnetic Lines Of Force : Imaginary lines used for convenience to designate the direction in which magnetic forces are acting as a result of magnetomotive force.

Magnetic Microphone : A microphone in which the sound waves vibrate a moving armature. The armature consists of a coil wound on the armature and located between the pole pieces of a permanent magnet. The armature is mechanically linked to the diaphragm.

Magnetic Poles : The section of a magnet where the flux lines are concentrated; also where they enter and leave the magnet.

Magnetic Susceptibility : The proportionality constant between the magnetisation M and the magnetic field strength H.

Magnetic Trip Element : A circuit breaker trip element that uses the increasing magnetic attraction of a coil with increased current to open the circuit.

Magnetism : The property possessed by certain materials by which these materials can exert mechanical force on neighbouring masses of magnetic materials and can cause currents to be induced in conducting bodies moving relative to the magnetised bodies.

Magnetisation : The total magnetic moment per unit volume of material. Also, a measure of the contribution to the magnetic flux by some material within an H field.

Magnetron Oscillator : An electron tube that provides a high power output. Theory of operation is based on interaction of electrons with the crossed electric and magnetic fields in a resonant cavity.

Magnitude Comparator : A circuit that compares one value with another. In digital logic a circuit that compares two binary numbers.

Manchester Data Encoding : Manchester encoding is a form of binary phase-shift keying (BPSK) that has gained wide acceptance as a modulation scheme for low-cost radio-frequency (RF) transmission of digital data. Its key characteristic is that it encodes data in a way that insures there will never be long strings of continuous zeros or ones. The guaranteed transitions means that the clock can be derived from the transmitted data, allowing the link to function with variable signal strengths from transmitters with imprecise, low-cost, data-rate clocks.

Margining : Margining is a test procedure that determines the "safety margin." A parameter is varied to determine the device's sensitivity or ability to perform given a range of inputs. A large number of parts can be characterised to determine a safe range for the specification, to guarantee performance and yield.

Mark : An interval during which a signal is present. Also the presence of an RF signal in cw keying. The key-closed condition (presence of data) in communications systems. A high level in an RS232 System [the opposite of a Space].

Martensite : A metastable Fe-C composition consisting of supersaturated carbon in iron that is the product of a diffusionless (athermal) transformation from austenite.

Mass Balance : An organised accounting of all inputs and outputs to an arbitrary but defined system. : Stated in other terms, the rate of mass accumulation within a system is equal to the rate of mass input less the rate of mass output plus the rate of mass generation within the system.

Mass : The SI unit of mass is the kilogram and is equal to the mass of the International Prototype kilogram kept at Sevres, Paris and is a cylinder of platinum and iridium.

Master Oscillator Power Amplifier (MOPA) : A transmitter in which the oscillator is isolated from the antenna by a power amplifier.

Master Oscillator : In a transmitter, the oscillator that establishes the carrier frequency of the output.

Matrix **:** The body constituent of a composite or two-phase alloy that completely surrounds the dispersed phase and gives the body its bulk form.

Max. DNL (LSB) **:** Maximum differential nonlinearly expressed in least significant bits.

Max. Hold Step (MV) **:** When switching between sample mode and hold mode, charge injection from stray capacitance causes the maximum voltage of the hold capacitor to change.

Maximum Contaminant Level (MCL) **:** The maximum allowable concentration of a given constituent in potable water.

Maximum Power Dissipation **:** The maximum total power dissipation a component will operate at within specification.

Maximum Usable Frequency **:** Maximum frequency that can be used for communications between two locations for a given time of day and a given angle of incidence.

Maxterm **:** A sum term containing every variable once and only once.

Mayer Method **:** A method of evaluating the propagation delay of a transmission line.

MBB (Make-before-break) **:** In a switching device, a configuration in which the new connection path is established before the previous contacts are opened. This prevents the switched path from ever seeing an open circuit. Applies to mechanical systems (e.g. that use relays or manual switches) and to solid-state analog multiplexers and switches.

MCM **:** (1) Multi-Chip Module (MCM): An integrated circuit package that contains two or more interconnected chips. (2)MCM is an abbreviation for thousands of circular mils, an old measurement of wire gauge. 1 MCM = 1 kcmil = 0.5067 square milimeters. A mil is 1/1000 inch. A wire 200 mils in diameter is 40 MCM. MCM is generally used for very large-diameter wire. Most wire uses AWG.

Mcps **:** (1) Megacycles per second (obsolete): Megahertz. (2) Megachips per Second (Mcps): In a Direct-Sequence Spread Spectrum signal, a "chip" is an encoding element.

Mcps is a measure of the speed at which chips can be generated by a circuit.

Media Independent Interface : A parallel digital bus used for 10Mbps and 100Mbps Ethernet.

MegaBaud : RS-232 logic-level compatible data rates that are 1Mbps or higher.

Melting Point : The temperature at which a solid substance changes to a liquid state.

Memory : A system for storing data or program instructions in binary form either permanently (ROM), semi-permanently (EPROM), or temporarily (RAM). Other methods also available.

MEMS : Acronym for Acronym for "Micro Electronic Mechanical Systems," or microelectromechanical systems: Systems that combine mechanical and electrical components and are fabricated using semiconductor fabrication techniques. Common examples are pressure and acceleration sensors which combine the sensor and amplification or conditioning circuitry. Other applications include switches, valves, and waveguides.

Mer : The group of atoms that constitutes a polymer chain repeat unit.

Mercuric Sulphide : The ore of mercury. Occuring as red crystals.

MESFET : A Metal-Semiconductor Field-Effect-Transistor uses a metal-semiconductor (Schottky) junction to create the conductive channel, rather than using a p-n junction as a JFET does; or a metal-oxide-semiconductor layer as a MOSFET uses.

Metabolism : The processes which sustain an organism, including energy production, synthesis of proteins for repair and replication.

Metal Oxide Varistor : A Metal Oxide Varistor (MOV, or surge-suppressor) is a discrete electronic component that diverts excessive voltage to the ground and/or neutral lines.

Metal : The electroposite elements and alloys based on these elements.

Metal Electrode Leadless Face : Surface Mount Components that have metallized terminals at each end of a cylindrical body. MELF components are designed to fit the same footprints as flat components, i.e.: 0805 (.08" x.05") and the 0603 (.06" x.03"). MELF devices may be Fuses, Resistors, or Diodes.

Metallic Armor : A protective covering for wires or cables. Made as a woven wire braid, metal tape, or interlocking metal cover. Types include wire braid, steel tape and wire armor. Made from steel, copper, bronze, or aluminum.

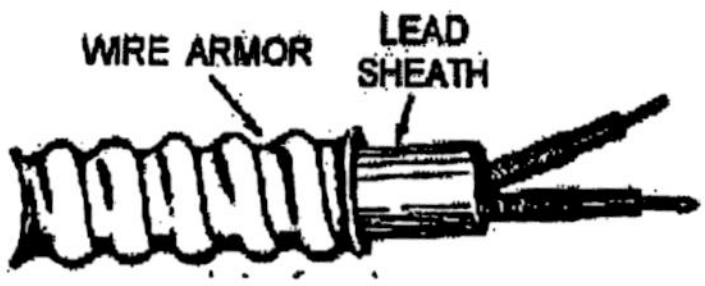

Metallic Armor

Metallic, Insulator : A shorted quarter-wave section of transmission line.

Metallic Rectifier : A metal-to-semiconductor, large area, contact device in which a semiconductor is sandwiched between two metal plates. This asymmetrical construction permits current to flow more readily in one direction than the other.

Meter : A device used to measure a specific quantity, such as current, voltage, or frequency. A panel mount analog meter is shown. However a digital meter may be more common these days.

Meteorology : The study of the atmosphere and weather of the lower atmosphere, below 100 km.

Metastability : A circuit condition that could cause a digital logic IC to reside in an invalid state.

Meter Movement : The part of the meter that moves to indicate some value. Obsolete reference, all equipment is digital, there are no analog meters being used, unless legacy units which have not been discarded.

Meter Shunt : A resistor placed in parallel with the meter terminals; used to provide increased range capability. Obsolete reference; a meter shunt was only required with analog meters which are in decline.

Mezzanine : The termed used to describe a daughter card which plugs into another card [motherboard] which is then plugged into a backplane.

MHz : Megahertz (MHz): Measurement of frequency million cycles per second.

Microcircuit : A circuit having high equivalent-circuit-element density, considered as a single part composed of interconnected elements on or within a single substrate to perform an electronic-circuit function.

Microcircuit Module : An assembly of microcircuits or a combination of microcircuits and discrete components that perform one or more distinct functions.

Microelectronics : The solid-state concept of electronics in which compact semiconductor materials are designed to function as an entire circuit or sub-assembly rather than as circuit components [ICs].

MicroLAN : A 1-Wire network. A low-cost network in which PCs or microcontrollers communicate digitally over twisted-pair cable using 1-Wire components.

MicroMonitor : A device that monitors three conditions vital to processor-controlled systems: power supply, software execution, and external override.

Microphone : An energy converter that changes sound energy into electrical energy.

Microprocessor : A very complex type of microchip which can manipulate data in binary form and carry out functions at high speed according to a programmed sequence of instructions stored in a memory device such as an EPROM.

Microprocessor Supervisor : A device that monitors a host microprocessor or microcontroller's supply voltage and, in some cases, its activity. It monitors for a fault condition and takes appropriate action, usually issuing a reset to the microprocessor.

Microstrip : Traces that are adjacent to a continuous reference plane. In a Printed Circuit Board [PCB] stackup, the microstrips are on the primary and secondary layers.

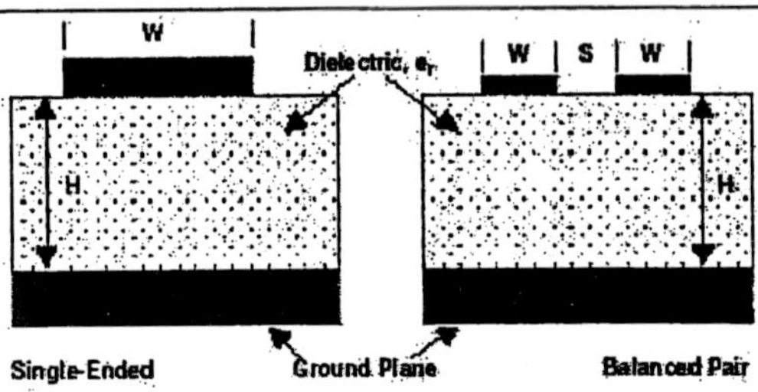

Microstrip Traces

Microstructure : The structural features of an alloy that are subject to observation under a microscope.

Miller Indices : A set of three integers that designate crystallographic planes, as determined from reciprocals of fractional axial intercepts.

Miller-Bravis Indices : A set of four integers that designate crystallographic planes in hexagonal crystals.

Milling Machine : Machine tool used to machine materials where the workpiece is fixed to a table and moved in XYZ relative to a rotating cutting tool. They exist in two basic forms: horizontal and vertical, which terms refer to the orientation of the cutting tool spindle.

MIMO : A Multiple Input, Multiple Output (MIMO) system has multiple antennas and multiple radios. It takes advantage of multipath effects, where a transmitted signal arrives at the receiver through a number of different paths. Each path can have a different time delay, and the result is that multiple instances of a single transmitted symbol arrive at the receiver at different times. Usually multipath is a source of interference, but MIMO systems use the fact that data will arrive at the receiver at different times through different paths to improve the quality of the data link.

Min LOS Sens : : The minimum sensitivity attainable with a programmable loss-of-signal feature.

Min Stable Closed Loop Gain : The minimum closed-loop gain for which the amplifer is stable.

Mismatch Loss : The loss of power delivered into a load as a result of the interconnection of devices having unequal impedances

Mixed Dislocation : A dislocation that has both edge and screw components.

Mixed Liquor Suspended Solids (MLSS) : The total suspended solids concentration in the activated sludge tank.

Mixed Liquor Volatile Suspended Solids (MLVSS) : The volatile suspended solids concentration in the activated sludge tank.

MMT (Methylcyclopentadienyl Manganese Tricarbonyl) : An octane boosting additive which also gives some protection to valve seats so was adopted as a lead substitute. Now known to be a neurotoxin with suspected links to the BSE / CJD crisis and hence has been banned in many countries.

Modem (Modulator/Demodulator) : An interface unit which makes it possible for a computer to communicate with other devices by serial link over long distances via a telephone line.

Mode Shifting : In a magnetron, the inadvertent shifting from one mode to another during a pulse.

Mode Skipping : Operation in which the magnetron fires randomly, rather than firing on each successive pulse as desired.

Modular Circuitry : A technique where printed circuit boards are stacked and connected together to form a module.

Modular Packaging : Circuit assemblies or sub-assemblies packaged to be easily removed for maintenance or repair.

Modulated Wave : A complex wave consisting of a carrier and a modulating wave that is transmitted through space.

Modulating Wave : An information wave representing intelligence.

Modulation : The process of impressing intelligence upon a transmission medium, such as radio waves.

Modulation Factor (M) : An indication of relative magnitudes of the RF carrier and the modulating signal.

Modulation Index : The ratio of frequency deviation to the frequency of the modulating signal.

Modulator : A device that produces modulation; that is, a device that varies the amplitude, frequency, or phase of an ac signal. A circuit used in servo-systems to convert a dc signal to an ac signal. The output ac signal is a sine wave at the frequency of the ac reference voltage. The amplitude of the output is directly related to the amplitude of the dc input.

Modulator Switching Device : Controls the on (discharge) and off (charge) time of the modulator.

Module : A circuit or portion of a circuit packaged as a removable unit. A separable unit in a packaging scheme displaying regularity of dimensions.

Modulus of Elasticity : The ratio of stress to strain for a material under perfectly elastic deformation.

Monel Metal : Strong and corrosion resistant. Applications include: turbine blades, propellers, pump fittings, condenser tubes.

Mono Jack : A jack designed to handle one channel of sound. There are many different sizes and styles of mono jacks, one example is shown below.

Monolithic Circuit : A circuit where all elements (resistors, transistors, and so forth) associated with the circuit are fabricated in-separably within a continuous piece of material (called the substrate), usually silicon.

Monolithic IC : ICs that are formed completely within a semiconductor substrate. Silicon chips.

Monostable Multivibrator : A multivibrator that has one steady state. A signal (trigger) must be applied to cause change of states.

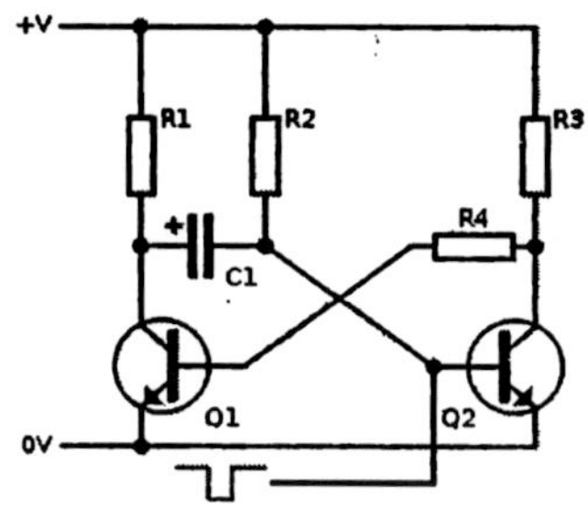

Monostable Multivibrator

Monotonic : A sequence increases monotonically if for every n, Pn + 1 is greater than or equal to Pn. Similarly, a sequence decreases monotonically if for every n, Pn + 1 is less than or equal to Pn. In plain language, the value rises and never falls; or it falls and never rises.

Morse Code : An out-dated form of communication using 'dot' and 'dashes'.

Mosfet : A semiconductor device that contains diffused source and drain regions on either side of a P- or N-channel area. Also contains a gate insulated from the channel area by silicon-oxide.

MOSFET : Metal-oxide semiconductor field-effect transistor; metal-oxide silicon field-effect transmitter. In a MOSFET, the conductive channel between the drain and source contacts is controlled by a metal gate separated from the channel by a very thin insulating layer of oxide. The gate voltage establishes a field that allows or blocks current flow.

MOSI (Master Out Slave In) : One of the four Serial Peripheral Interface (SPI) pins. Most commonly applied to make a light-emitting device, such as a display, match the human eye's brightness curve. In other terms: A gamma correction function can be used to alter the luminance (light intensity) of a display such that its brightness (the human-perceived values) looks correct.

Most Significant Digit (MSD) : The MSD is the digit whose position within a given number expression has the greatest weighting power.

Motor : A machine that converts electrical energy to mechanical energy. It is activated by ac or dc voltage, depending on the design.

Electrical Motor

Motor Load : Any device driven by a motor. Typical loads are drills, saws, water pumps, rotating antennas, generators, and so forth. The speed and power capabilities of a motor must be matched to the speed and power capabilities of the motor load.

Motor Reaction : The force created by generator armature current that tends to oppose the normal rotation of the armature.

Motor Starters : Large resistive devices placed in series with dc motor armatures to prevent the armature from drawing excessive current until armature speed develops counter emf. The resistance is gradually removed from the circuit either automatically or manually as motor speed increases.

Mounting Hardware : Any hardware required to secure a device or component to another device in the system.

MOV : Metal Oxide Varistor. A resistor that changes value with applied voltage. A varistor may also be called a VDR

[Voltage Dependent Resistor].Varistors will have a negative voltage coefficient. MOV devices are used in parallel with the load.

Moving-Vane Meter Movement : A meter movement that uses the magnetic repulsion of the like poles created in two iron vanes by current through a coil of wire; most commonly used movement for ac meters. This is an Obsolete terms, most meters being produce are digital, there is no meter movement.

MSA (Measurement Systems Analysis) : MSA is a method for ensuring product test measurements are reliable, robust, and of good statistical merit.

MSB (Most-significant bit) : In a binary number, the MSB is the most weighted bit in the number. Typically, binary numbers are written with the MSB in the left-most position; the LSB is the furthest-right bit.

Msps (Megasamples per second) : A measure of speed in digitising systems, samples per second dictates the maximum frequencies that can be accurately captured.

MTBE (Methyl Tertiary Butyl Ether) : An oxygenated gasoline additive with a low volatility used in reformulated gasoline and also in racing fuels because of its high octane rating of 117 RON. Became a mandatory fuel component in some US States until it was found to significantly pollute ground water because of its exceptional ability to exploit any leak from storage tanks, so it has now been outlawed in California and other places.

Multiconductor : More than one conductor, as in a cable. Two or more insulated conductors (solid or stranded) contained in a common covering or jacket.

Multicouplers : Couplers that patch receivers or transmitters to antennas. They also filter out harmonics and spurious responses and impedance-match the equipment.

Multielectrode Tube : An electron tube normally classified according to its number of electrodes (the multi-electrode tube contains more than three electrodes).

Multielement Array : An array that consists of one or more arrays and is classified as to directivity.

Multielement Parasitic Array : An array that contains two or more parasitic elements and a driven element.

Multiloop Servosystem : A servo-system that contains more than one servo loop; each loop is designed to perform its own function.

Multimeter : A single meter combining the functions of an ammeter, a voltmeter, and an ohmmeter.

Multimeter

Multipath : In radio transmission, multipath refers to the simultaneous reception of two copies of the signal, that arrive via separate paths with different delays. A common example is when a signal bounces off a building or other object and is received along with the direct (unbounced) signal. In television reception, this causes "ghosting" one sees a faded echo on the screen horizontally displaced from the main image.

Multiplex : (1) Combining two signals (which can be analog or a digital stream) into one in such a way that they can later be separated. Examples are OFDM; standard FM stereo broadcast (in which left and right are multiplexed onto one base band signal); standard television in which video and several audio signals shared the channel; and time-division multiplexing which gives each signal a separate time-slice. (2) An array of analog switches, usually on a single CMOS chip, that allows one input signal to be routed to any of several output lines, depending on the value of a set of digital control lines. A multiplexer can also be used in the opposite direction, allowing the array to connect one of several input lines to the output, depending on the control lines. Several of these can be implemented on one chip to make a multi-channel version.

Multiplexing : A method for simultaneous transmission of two or more signals over a common carrier wave. The T1 Interface uses multiplexing to transmit data over trunk lines. An IC circuit that transmits one of two signals from the input to the output, refer to IC Multiplexers. Multiplexing is used when driving many single channels to a remote location, but installing a separate line for each channel would be prohibitive.

Multiplication Factor : The number of times an input frequency is multiplied.

Multipoint Injection : A system employing a separate injector (or multiple injectors) for each cylinder.

Multispeed Synchro Systems : Systems that transmit data at different transmission speeds; for example, dual-speed and tri-speed synchro systems.

Multi-Unit Tube : An electron tube containing two or more units within the same envelope. The multi-unit tube is capable of operating as a single-unit tube, or each unit can operate as a separate tube.

Multivibrator : A form of relaxation oscillator which comprises two stages that are coupled so that the input of one is derived from the output of the other.

Multivibrator Modulator : An astable multivibrator used to provide frequency modulation. The modulating af voltage is inserted in series with the base return of the multivibrator transistors to produce the frequency modulation.

Mutual Flux : The total flux in the core of a transformer that is common to both the primary and secondary windings. The flux links both windings.

Mutual Inductance : A circuit property existing when the relative position of two inductors causes the magnetic lines of force from one to link with the turns of the other. The symbol for mutual inductance is M.

mV : A millivolt (mV) is 1/1000 of a volt.

N

NAND Circuit : A combination of a NOT function and an AND function in a binary circuit that has two or more inputs and one output. The output is logic 0 only if ALL inputs are logic 1; it is logic 1 if ANY input is logic 0.

Nanovolt : ***Nanovolt (nV)***: Unit of measure. A billionth of a volt.

N-Channel Fet : A type of Field Effect Transistor, which is voltage controlled.

Near End Crosstalk Ratio : Crosstalk measured at the transmit end of a transmission line.

Near End : The originating side of a circuit. The opposite of Far End [terminating side of a circuit].

Near Field : The electromagnetic field that exists within one wavelength of a source of radiation.

Near Synchronous Orbit : An orbit in which the satellite rotates close to but not exactly at the same speed as the earth.

Negation : The process of inverting the value of a function or variable.

Negative Alternation : That part of a sine wave that is below the reference level.

negative Clamper : A circuit that clamps the upper extremity of the output waveshape to a dc potential of 0 volts.

Negative Electrode : A terminal or electrode having more electrons than normal. Electrons flow out of the negative terminal of a voltage source.

Negative Feedback : Feedback in which the feedback signal is out of phase with the input signal. Also called DEGENERATIVE FEEDBACK.

Negative Logic : The form of logic in which the more positive voltage level represents a logic 0, FALSE, or LOW and the more negative voltage represents a logic 1, TRUE, or HIGH.

Negative Temperature Coefficient : A characteristic of a semiconductor material, such as silver sulfide, in which resistance to electrical current flow decreases as temperature increases.

Negative-Resistance Element : A component having an operating region in which an increase in the applied voltage increases the resistance and produces a proportional decrease in current. Examples include tunnel diodes and silicon unijunction transistors.

Neper : A logarithmic unit used for expressing the ratio of two voltages.

Network Analyser : A measuring instrument for evaluating the complex characteristics of linear electrical circuits.

Network Polymer : A polymer composed of trifunctional mer units that form three-dimensional molecules.

Network : A combination of electrical components. In a parallel circuit it is composed of two or more branches.

Network Topologies : A Network is any connection of two or more computers that enables them to communicate. Networks may include transmission devices, servers, cables, routers and satellites. The phone network is the total infrastructure for transmitting phone messages.

Neutral : In a normal condition, hence neither positive nor negative. A neutral object has a normal number of electrons (the same number as protons). A current-carrying conductor normally tied to ground so that the electrical potential is zero.

Neutralisation : The process of counteracting or "neutralising" the effects of inter-electrode capacitance.

Newton's Laws of Motion : *Newton's First Law*: Every body continues in its state of rest or of uniform motion in a straight line unless acted on by some external force : *Newton's Second Law:* The rate of change of momentum of a body is directly proportional to the external force acting on the body and takes place in the direction of the force. *Newton's Third Law:* If body A exerts a force on body B, then B exerts an equal and opposite force on A.

Nitrification : The biological oxidation of ammonia and ammonium sequentially to nitrite and then nitrate.It occurs naturally in surface waters, and can be engineered in wastewater treatment systems. The purpose of nitrification in wastewater treatment systems is a reduction in the oxygen demand resulting from the ammonia.

Nitrogenous Oxygen Demand (NOD) : The amount of oxygen required to oxidise any ammonia present in a water.

NMOS : An n-channel metal-oxide semiconductor (nMOS) transistor is one in which n-type dopants are used in the gate region (the "channel"). A positive voltage on the gate turns the device on.

Noncrystalline : The solid state wherein there is no long-range atomic order. Sometimes used synonymously with the terms amorphous, glassy and vitreous.

Non-inverting Amplifier : An amplifier that does not invert the incoming signal at the output. The circuit example shows a Non-inverting Op Amp

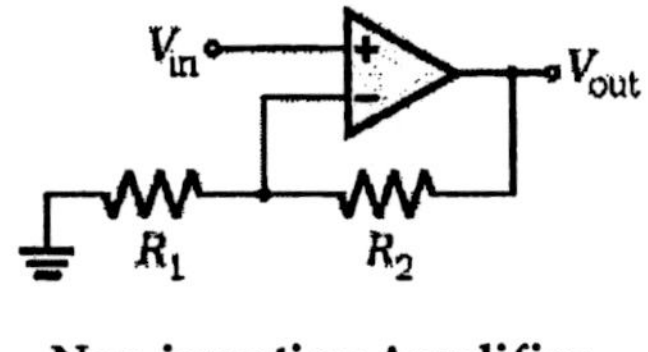

Non-inverting Amplifier

Nonpoint Source pollution (NPSP) : Any pollution from a source which cannot be attributed to a particular discharge point, e.g. from agricultural crops, city streets, construction sites, etc.

Nonvolatile : Nonvolatile (NV) RAM is memory which retains its stored value when power is removed.

NOR : A logic function of A and B that is true if both A and B are false.

NOR Gate : An OR gate that is followed by an inverter to form a binary circuit in which the output is a logic 0 if ANY of the inputs is a logic 1 and the output is a logic 1 only if all the inputs are a logic 0.

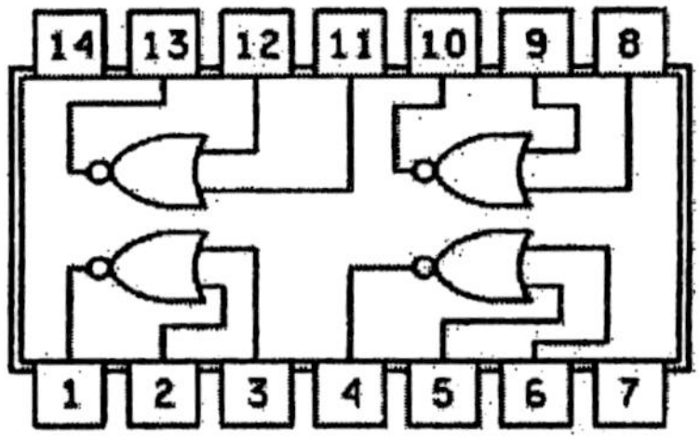

NOR Gate IC Pin-Out

Normal : The imaginary line perpendicular to the point at which the incident wave strikes the reflecting surface. Also called the perpendicular.

Not Circuit : A binary circuit with a single output that is always the opposite of the input. Also called an Inverter Circuit.

NPN : An NPN transistor is formed by introducing a thin region of P-type material between two regions of N-type material

Noxious Fumes : A combination of inert and corrosive gases usually associated with exhaust fumes or industrial by-products gases which can cause corrosive effects on temperature and pressure sensors when exposed.

NRZ (Non Return to Zero) : A binary encoding scheme in which ones and zeroes are represented by opposite and alternating high and low voltages, and where there is no return to a zero (reference) voltage between encoded bits. That is, the stream has only two values: low and high.

NTSC : NTSC is the color television standard established by the National Television Standards Committee in the United States in 1953. The NTSC standard's distinguishing feature was that it added color to the original 1941 black and white television standard in such a way that black and white TVs continued to work. (

Nucleation : The initial stage in a phase transformation. It is evidenced by the formation of small particles (nuclei) of the new phase, which are capable of growing.

Null : A balanced condition of a circuit that represents zero output, as in Off-Set Null of an Op Amp. The minimum output amplitude of a circuit. On a polar-coordinate graph, the area that represents minimum or 0 radiation.

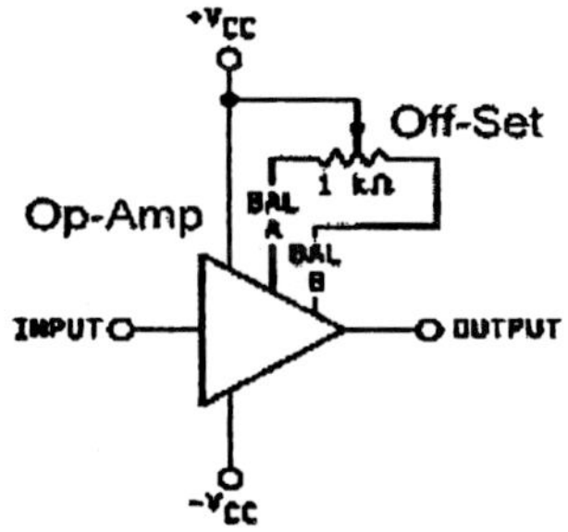

Off-Set Null

Number : A mathematical entity that may indicate quantity or amount of units. An abstract mathematical symbol for expressing a quantity

Number Drills : This system of numbering drills is no longer used, but may be of use in deciphering old drawings.

Number System : A number representation system. Any system for the representation of numbers

Numeral : A discrete representation of a number. For example, twelve, 12, XII, are three different numerals that represent the same number. A numeric word that represents a number.

Nyquist : In A/D conversion, the Nyquist principle (derived from the Nyquist-Shannon sampling theorem) states that the sampling rate must be at least twice the maximum bandwidth of the analog signal in order to allow the signal to be reproduced. The maximum bandwidth of the signal (half the sampling rate) is commonly called the Nyquist frequency (or Shannon sampling frequency).

O

Obsolete : Products that have been replaced by a newer or more advanced function. Products that have been discontinued or no longer being manufactured.

Octahedral position : The void space among closed-packed, hard sphere atoms or ions for which there are six nearest neighbours. An octahedron is curcumscribed by lines constructed from centers of adjacent spheres.

Octal Number System : A number system based on powers of eight.

OFDM (Orthogonal Frequency Division Multiplexing) : A method for multiplexing signals which divides the available bandwidth into a series of frequencies known as tones. Flarion uses the 5GHz channel and divides each channel into 400 discrete tones (each at slightly different frequency). Orthogonal tones do not interfere with each other when the peak of one tone corresponds with the null. All frequencies fade but the rapid switching, frequency-hopping technique is intended to allow more robust data service.

Off-Line Test Equipment : Equipment that tests and isolates faults in modules or assemblies removed from systems.

Off-Set Null : A two terminal balance adjustment on an Op-Amp used to zero or null the output.

OHM : The unit of electrical resistance. That value of electrical resistance through which a constant potential difference of 1 volt across the resistance will maintain a current flow of 1 ampere through the resistance.

Ohm's Law : The current in an electrical circuit is directly proportional to the electromotive force in the circuit. The most common form of the law is E = IR, where E is the electromotive force or voltage across the circuit, I is the current flowing in the circuit, and R is the resistance of the circuit.

Ohm's Law : A voltage of 1V (volt) across a conductor with a resistance of 1W (ohm) causes a current of 1A (amp) to flow through it. V=IR. R=V/I. I=V/R.

Ohmmeter : A meter used to measure resistance.

Ohmmeter

Ohms Per Square : The resistance of any square area of thin film resistive material as measured between two parallel sides.

Oil Hardening Steel : Heat the steel to 770 to 800°C and soak at this temperature for 3/4 hour per inch (25.4mm) thickness of material. Quench in a bath of oil inserting the heaviest end of the piece first (this way relieves stresses and reduces distortion). Further heat treatment at 100 to 120°C for a short while and then allowing the piece to cool naturally in air will reduce the stresses further and the possibility of cracking.

Oilcan Tube : A type of planar tube, similar to the lighthouse tube, which has cooling fins. The oilcan tube is designed to handle large amounts of power at uhf frequencies.

OLED (Organic Light-Emitting Diode) : An LED made with organic materials. The diodes in displays made with OLEDs emit light when a voltage is applied to them. The pixel diodes are selectively turned on or off to form images on the screen. This kind of display can be brighter and more efficient than current LCD displays.

On Board Diagnostics (OBD) : Originally introduced by California Air Resources Board (CARB) in 1988, OBD1 required vehicle manufacturers to provide a system which could identify emission related malfunctions in electronic engine management systems and display a warning to the driver and a fault code system to provide a technician with guidance.

ONU : Optical Network Termination, also called ONU (Optical Network Unit), refer to the consumer end equipment in an optical Fiber to the Home (FTTH) link. The ONT/ONU receives downstream data from the OLT (Optical Line Termination) through the passive optical splitters and provides video, voice, and broadband services to the consumer.

Operational Amplifier : The ideal op amp is an amplifier with infinite input impedance, infinite open-loop gain, zero output impedance, infinite bandwidth, and zero noise. It has positive and negative inputs which allow circuits that use feedback to achieve a wide range of functions. Using op amps, it's easy to make amplifiers, comparators, log amps, filters, oscillators, data converters, level translators, references, and more.

Open Loop Operation : Usually used to describe a situation where a system with Lambda feedback switches to operation in which the Lambda signal is ignored so that fuelling values more appropriate to the conditions can be used without feedback correction. Full load enrichment is often provided by going 'open loop'.

Optimum Working Frequency : The most practical operating frequency that can be used with the least amount of problems and is roughly 85 percent of the maximum usable frequency.

Optoelectronic Devices : Devices that either produce or use light in their operation.

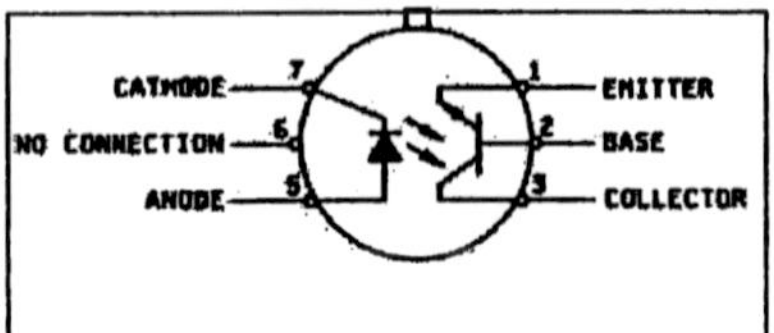

Opto Device

Opus Ignition : An early electronically triggered ignition named after its operating method 'oscillating pickup system'. Does not control coil current, therefore requires ballast resistor, but is capable of high spark rate. Used on racing engines (e.g. Cosworth DFV) and multi-cylinder production engines (e.g. early Jaguar V12).

OR : Combining two signals so that the output is on if either signal is present. This can be accomplished by an OR logic gate (two inputs, one output which is high if either input is). It can also be done with a "wired-OR" connection in which two signals are simply wired together and either one of them can raise the level. This works when the signals are driven by a source that only pulls up or only pulls down, with a resistive load (e.g. an "open collector" output).

Oscillator : An oscillator is a non-rotating device that produces alternating current. The frequency is determined by the characteristics of the device.

Out-of-Circuit Meter : A meter that is not permanently installed in a circuit. Usually portable and self-contained, these meters are used to check the operation of a circuit or to isolate troubles within a circuit.

Output End : The end of a transmission line that is opposite the source; receiving end.

Output Impedance : The impedance that is presented to the load by the transmission line and its source.

Output to Input Ratio : The ratio between the sensed current and the output current of the amplifier.

Overcurrent : Any current in excess of the rated current of equipment or the ampacity of a conductor. It may result from overload, short circuit, or ground fault. A current in excess of rating may be accommodated by certain equipment and conductors for a given set of conditions. Hence the rules for over-current protection are specific for particular situations.

Overdriven : When the input signal amplitude is increased to the point that the transistor goes into saturation and cutoff.

Overload : Operation of equipment in excess of normal, full load rating, or of a conductor in excess of rated ampacity which, when it persists for a sufficient length of time, would cause damage or dangerous over-heating. A fault, such as a short circuit or ground fault, is not an overload.

Overmodulation : A condition that exists when the peaks of the modulating signal are limited.

Over-Run Cut Off : Condition in which no fuel injection takes place if the throttle is fully closed above a predetermined engine speed - typically about 1300 r.p.m. The cutting in and cutting out speeds are separated by 2 - 300 r.p.m. to prevent oscillation between states.

Over-Run Valve : A spring loaded valve which opens to admit air into the inlet manifold plenum when a predetermined vacuum (typically 20 ins mercury) is reached under trailing throttle conditions to reduce exhaust emissions and prevent 'pop back'. Essential for engines which do not have over-run cut off. Some later engine management systems use the idle air valve to perform the same function.

Overvoltage Protection : Overvoltage Protector (OVP) refers to a circuit that protects downstream circuitry from damage due to excessive voltage. An OVP monitors the DC voltage coming from an external power source, such as an off-line power supply or a battery, and protects the rest of the connected circuitry using one of two methods: a crowbar clamp circuit or a series-connected switch. The crowbar short-circuits or clamps the supply line to limit the voltage, possibly triggering other forms of protection such as a fuse.

Oxidative phosphorylation : The synthesis of the energy storage compound adenosine triphosphate (ATP) from adenosine diphosphate (ADP) using a chemical substrate and molecular oxygen.

Oxygenated Compounds (Gasoline Blend Components): Comprising the alcohols Methanol, Ethanol, Iso-Propyl Alcohol (IPA), Tertiary Butyl Alcohol (TBA), and the ethers DIPE (Di-Iso-Propyl Ether), ETBE (Ethyl Tertiary Butyl Ether), MTBE (Methyl Tertiary Butyl Ether), TAME (Tertiary Amyl Methyl Ether). These are compounds of high octane rating and containing oxygen used as gasoline blend components and as lead substitutes.

P

PA (Power Amplifier) : An amplifier used to drive significant power levels. An audio amplifier that drives a loudspeaker and the final stage of a transmitter are common examples.

Packet Format : The structure of data, address, and control information in a packet.

Packet Switching : The process of routing and transferring data by means of addressed packets so that a channel is occupied during the transmission of the packet only, and upon completion of the transmission the channel is made available for the transfer of other traffic.

PAL (Phase Alternate Line) : A television standard used in most of Europe. Similar to NTSC, but uses subcarrier phase alternation to reduce the sensitivity to phase errors that would be displayed as color errors. Commonly used with 626-line, 50Hz scanning systems, with a subcarrier frequency of 4.43362MHz.

Parallax Error : The error in meter readings that results when you look at a meter from some position other than directly in line with the pointer and meter face. A mirror mounted on the meter face aids in eliminating parallax error.

Parallel Circuit : Two or more electrical devices connected to the same pair of terminals so separate currents flow through each; electrons have more than one path to travel from the negative to the positive terminal.

Parallel Interface : A parallel interface (as distinguished from a serial interface) is one in which data is sent on several wires (or several wireless channels) at once. Examples: GPIB, byte-wide parallel interfaces to data converters, memory and data buses on computer boards and backplanes. In contrast, a serial interface uses one wire or wire-pair or wireless channel (or one in each direction).

Parallel Port : A port through which two or more data bits are passed simultaneously, such as all the bits of an 8-bit byte, and that requires as many input channels as the number of bits that are to be handled simultaneously. Centronics, IEEE-1284

Parallel-Negative Limiter : A resistor and diode, connected in series with the input signal, in which the output is taken across the diode and the negative alternation is eliminated.

Parallel-Positive Limiter : A resistor and diode connected in series with the input signal, in which the output is taken across the diode and the positive alternation is eliminated.

Parallel-Resonant Circuit : A resonant circuit in which the source voltage is connected across a parallel circuit (formed by a capacitor and an inductor) to furnish a high impedance to the frequency at which the circuit is resonant.

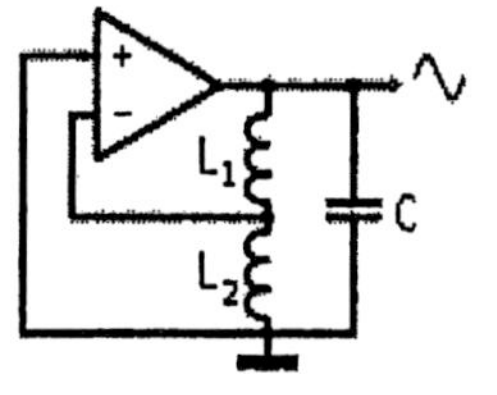

Oscillator

Paramagnetism : A relatively weak form of magnetism that results from the independent alignment of atomic dipoles (magnetic) with an applied magnetic field.

Parasite Power : The device derives its supply power directly from the serial interface (1-Wire).

Parity Check : The process of checking the accuracy of binary data, by counting the total number of 1s, or 0s, to insure that there is always odd or always even amount depending on the parity.

Parity : A method of checking the accuracy of binary data. In binary-coded data, a condition that is maintained such that, in any permissible coded expression, the total number of 1s, or 0s, is always odd or always even. A binary digit appended to a group of digits or character so that there are an even number of 1s or 0s in the stream, or an odd number of 1s or 0s in the data field.

Parity Error : A condition that occurs when the received number of 1s, or 0s do not equal the expected value. A signal output from a device that indicates the parity or number of 1s or 0s do not match the expected count.

Partition Locking : The ability to lockout writes and/or reads to certain sections of the memory.

Passive Filter : A filter that does not use active components.

Passive Satellite : A satellite that reflects radio signals back to earth.

Passive Serial Interface : A term used with programming an FPGA. In the Passive Serial [PS] interface, an external host (configuration device, embedded processor, or host PC) controls configuration of the FPGA. Also called Passive Serial Mode, or Passive Serial Scheme; where Passive Serial may be Abbreviated as PS.

Patch Panel : A Panel used to tie different circuitry together by means of Patch Cords, usually of the same or similar type, appearing on jacks for monitoring, interconnecting, and testing purposes. Patch Panels facilitate flexibility in the use, routing or restoration of a variety of circuit types. There are a number of standard patch panels; Bantam, 1/4" Longframe, Mini WECO, XLR Patchbays, Telecom Patchbays. Also called a Patch Bay, Although sometimes a patch panel is referred to as a part of a patch bay.

pC **:** (1) pC: Picocoulomb(s), a unit of electrical charge. (2) PC: Printed circuit. (3) PC: Personal Computer.

PC Card **:** Add-in cards that conform to the PC Card specification (formerly called PCMCIA). A PC Card is a removable device, approximately the size of a credit card, designed to plug into a matching slot.

PCI (Peripheral Component Interconnect) **:** A standard interface used primarily on computer backplanes to connect interface cards and peripheral devices to the processor bus. PCI is often used for video display cards, network interfaces (e.g. Ethernet), and peripheral interfaces such as SCSI or USB.

PCI Express **:** PCI Express® (Peripheral Component Interconnect Express), officially abbreviated as PCIe®, is a computer expansion card standard designed to replace the older PCI, PCI-X, and AGP standards. It is used to link motherboard-mounted peripherals and as an expansion card interface for add-in boards. The PCIe electrical interface is also used in a variety of other standards, most notably the ExpressCard laptop expansion card interface.

PCM **:** Pulse-Code Modulation (PCM) is the conversion of an analog signal (e.g. audio) into digital, binary (0 or 1), coded pulses, decreasing noise susceptibility. PAM, PFM and PWM are examples of PCM methods.

PCMCIA (Personal Computer Memory Card International Association) **:** A standard for miniaturised laptop expansion cards for modems, storage, and other devices. The standard was officially renamed "PC card."

PCS (Personal Communications Service) **:** An American generic term for a mass-market mobile phone service, emphasising personal communication, independent of the technology used to provide it. PCS includes such digital cellular technologies as GSM 1900, CDMA and TDMA IS-136. 2G, CDMA, Digital, GSM, TDMA.

PDC (Personal Digital Cellular) **:** The digital wireless standard used in Japan. PDC uses TDMA air interface.

Peak Inverse Voltage : Peak Inverse Voltage (PIV) or Peak Reverse Voltage (PRV) refer to the maximum voltage a diode or other device can withstand in the reverse-biased direction before breakdown. Also may be called Reverse Breakdown Voltage. Note that PIV is also an abbreviation for FIPS 201 Personal Identity Verification.

Pearlite : A two-phase microstructure found in some steels and cast irons. It results from the transformation of austenite of eutectoid compositions and consists of alternating layers of alpha-ferrite and cementite.

Permittivity : The proportionality constant between the dielectric displacement D and the electric field E.

Pewter : Traditionally an alloy containing 80% tin and 20% lead.

PFM (Pulse-Frequency Modulation) : A pulse modulation technique in which the frequency is varied with the input signal amplitude. The duty cycle of the modulated signal does not change. Because it is always a square wave with changing frequency, PFM is also referred to as square-wave FM.

PFMEA (Process Failure Mode and Effects Analysis) : A methodology for assessing the weaknesses of production processes and the potential effects of process failures on the product being produced.

Phase : The angular relationship between two alternating currents or voltages when the voltage or current is plotted as a function of time. When the two are in phase, the angle is zero; both reach their peak simultaneously. When out of phase, one will lead or lag the other; that is, at the instant when one is at its peak, the other will not be at peak value and (depending on the phase angle) may differ in polarity as well as magnitude.

Phase Angle : The number of electrical degrees of lead or lag between the voltage and current waveforms in an ac circuit.

Phase Diagram : A graphical representation of the relationships between environmental constraints, composition, and regions of phase stability, ordinarily under conditions of equilibrium.

Phase Modulation : [PM] Angle modulation in which the phase of the carrier is controlled by the modulating waveform. The amplitude of the modulating wave determines the amount of phase shift, and the frequency of the modulation determines how often the phase shifts.

Phase Noise : The rapid short-term, random fluctuations in the phase of a wave.

Phase Shifter : A device used to change the phase relationship between two ac signals.

Phase Splitter : A device that provides two output signals from a single input signal. The two output signals differ from each other in phase (usually by 180 degrees).

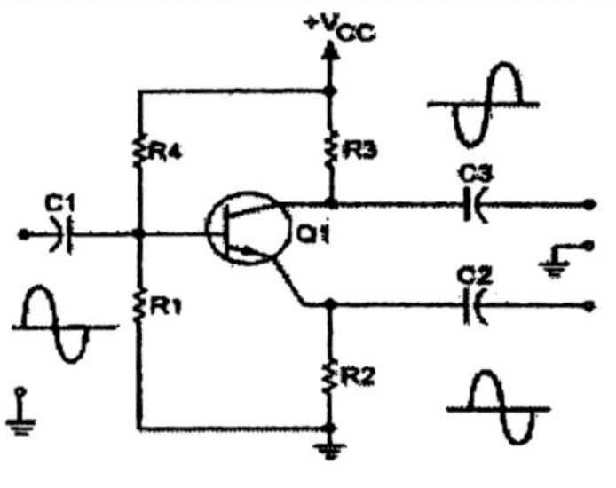

Phase Splitter

Phase Transformation : A change in the number and/or character of the phases that constitute the microstructure of an alloy.

Phase Voltage : The greatest root mean square (effective) difference of potential between any two legs of the circuit.

Phase-Locked Loop : [PLL] An electronic circuit that controls an oscillator so that it maintains a constant phase angle relative to a reference signal. An electronic circuit with a voltage or current-driven oscillator that is constantly adjusted to match in phase (and thus lock on) the frequency of an input signal.

Phase-Shift Keying : Similar to ON-OFF cw keying in AM systems and frequency-shift keying in FM systems. Each time a mark is received, the phase is reversed. No phase reversal takes place when a space is received.

Photocell : A light-controlled variable resistor which has a light-to-dark resistance ratio of 1:1000.

PhotochemicalP: Chemicals which react photochemically (in the presence of sunlight) to destroy ozone in the stratosphere.

Photodiode : A light-controlled PN junction. Current flow increases when the PN junction is exposed to an external light source. A semiconductor diode that produces, as a result of the absorption of photons, (a) a photovoltage or (b) free carriers that support the conduction of photocurrent. Photodiodes are used for the detection of optical communication signals and for the conversion of optical power to electrical power.

Photoelectric Voltage : A voltage produced by light.

Photoetching : Chemical process of removing unwanted material in producing printed circuit boards.

Photon : A quantum unit of electromagnetic energy.

Photophosphorylation : The synthesis of the energy storage compound adenosine triphosphate (ATP) from adenosine diphosphate (ADP) using solar energy.

Phototransistor : An optoelectronic device that conducts current when exposed to light. Produces more current and is much more sensitive to light than the photodiode.

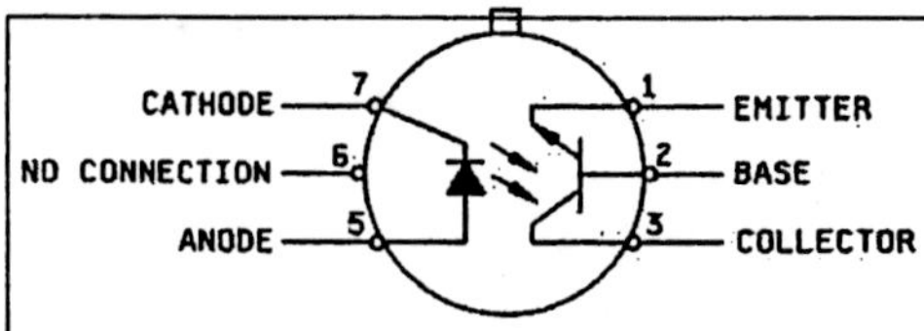

Phototransistor

Photovoltaic Cell (Solar Cell) : A device that acts much like a battery when exposed to light and converts light energy into electrical energy.

Physical Block Address : [PBA] The PBA is the fixed, physical address of a block of data on the flash drive.

pi : A mathematical constant whose value is the ratio of any circle's circumference to its diameter.

pi Filter : A filter circuit whose passive elements are arranged in the form of the mathematical symbol for pi.

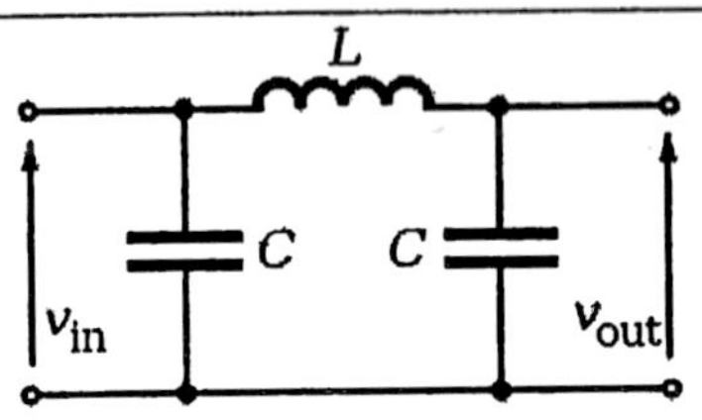

Low Pass pi Filter

Pickup - Hall Effect Type : A device widely used to produce trigger or reference signals from rotating shafts, often used inside distributors. A magnetically sensitive transistor switches as it detects the passage of a magnetic pole. Usually has three wires arranged in order: 5v, signal output, 0v

Pickup - Magnetic or Inductive Type : A very common sensor used to detect the passage of metal teeth or pegs from the changes caused in a magnetic field. Produces an alternating voltage swinging positive to negative (or vice versa) as each tooth passes, the magnitude increasing with speed. Usually has two connecting wires which are often screened against interference. Widely used to generate speed or register signals from distributors, crankshafts, camshafts and ABS wheel speed sensors.

Piezoelectric : A dielectric material in which polarisation is induced by the application of external forces.

Pin Electronics : Electronic circuitry in an automated tester (ATE system) that connects to the device under test. Pin electronics can deliver signals, power, or precise voltages and currents, and can measure the pin's response, drive, and electrical characteristics.

Pine : Easily worked and used where moderate strength and durability are required. Turpentine is obtained from the pine.

Pinking (also Pinging) : Descriptive name for relatively harmless combustion noise caused by over-advanced ignition for the fuel in use, usually occurring at low speed and high load. Not to be confused with detonation or pre-ignition.

Pipe Bending : There are many situations where you want to bend a pipe, but if you do not support it correctly the pipe walls will collapse.

PKI (Public Key Infrastructure) : A combination of standards, protocols, and software that creates, edits, and revokes digital public key certificates.

Planck's Constant : A universal constant that has a value of 6.63 x 10-34 J.

Plastic : A solid material in the primary ingredient of which is an organic polymer of high molecular weight.

Plastic Deformation : Deformation that is permanent or non recoverable after release of the applied load.

Plasticizer : A low molecular weight polymer additive that enhances flexibility and workability and reduces stiffness and brittleness.

PLC (Programmable Logic Controller) : PLC is a ruggedised, microprocessor-based system which provides factory or plant automation by monitoring sensors and controlling actuators in real time.

Plenum Chamber : A relatively large volume air chamber from which the inlet tracts run to the engine ports. A single throttle is usually used to control airflow into the plenum. The various vacuum signal tappings, idle control valves, over-run valves, etc. Are generally connected to the plenum.

Plesiochronous Digital Hierarchy : The time-division multiplexed network used by telecommunications companies to transport phone calls and data over copper cabling. The entire network shares a common frequency throughout it's tree-like structure, although phase and time delay variations exists at various points along the edge of the network.

PLL : A Phase-locked Loop is a control system that generates a signal that has a fixed relation to the phase of a "reference" signal. A phase-locked loop circuit responds to both the frequency and the phase of the input signals, automatically raising or lowering the frequency of a controlled oscillator until it is matched to the reference in both frequency and phase.

PMIC : Power Management Integrated Circuit. Circuits used to regulate and control power.

pMOS : A p-channel metal-oxide semiconductor (pMOS) transistor is one in which p-type dopants are used in the gate region (the "channel"). A negative voltage on the gate turns the device on.

PMR (Private Mobile Radio) : Radio bands generally for use within a defined user group, such as the emergency services or by the employees of a mining project.

PoE (Power-over-Ethernet) : A means for delivering power to a remote device using the same cable lines used to deliver Ethernet data.

Point Defect : A crystalline defect associated with one or, at most, several atomic sites.

Point-of-Load : Point-of-load (POL) power supplies solve the challenge of high peak current demands and low noise margins, required by high-performance semiconductors such as microcontrollers or ASICs, by placing individual power supply regulators (linear or DC-DC) close to their point of use.

Poisson's Ratio : For elastic deformation, the negative ratio of lateral and axial strains that result from an applied axial stress.

Polar Molecule : A molecule in which there exists a permanent electric dipole moment by virtue of the asymmetrical distribution of positively and negatively charged regions.

Polarisation (electronic) : For an atom, the displacement of the center of the negatively charged electron cloud relative to the positive nucleus, which is induced by an electric field.

Polarisation (ionic) : Polarisation as a result of the displacement of anions and cations in opposite directions.

Polarisation (orientation) : Polarisation resulting from the alignment (by rotation) of permanent electric dipole moments with an applied electric field.

Polycrystalline : Referring to crystalline materials that are composed of more than one crystal or grain.

Polymer : A solid, nonmetallic (normally organic) compound of high molecular weight the structure of which is composed of small repeat (or mer) units.

Polymorphism : The ability of a solid material to exist in more than one form or crystal structure.

PON (Passive optical network) : A cost-effective way to provide high performance Fiber to the Home (FTTH) connectivity via shared optical fiber. PON connects up to 32 (or more) homes on the same network using passive optical components (splitters).

Potentiometer : A special adaption of a resistor in which a sliding contact can be moved along the length of the resistor to tap off the voltage at any point. Widely used to sense throttle angle. If there is 5 volts across the potentiometer and the slider is one quarter travel from the lower end it would see 1.25 volts, however potentiometers can also be obtained with non linear scales, but throttle potentiometers are rarely, if ever, of that type.

Potentiometer

Power Added Efficiency : In an RF power amplifier, power added efficiency (PAE) is defined as the ratio of the difference of the output and input signal power to the DC power consumed. In other words: PAE = (PRFOUT - PRFIN)/PDC = (PRFOUT - PRFIN)/(VDC*IDC)

Power Fail : A feature in a microprocessor supervisory circuit that provides early warning to the microprocessor of imminent power failure.

PowerCap : A special surface-mount package with access to the internal cavity via an openable top. This packaging scheme allows easy upgrade of NV RAMs without having to change the PCB hardware layout. The user can simply open the lid and swap out the IC.

PPAP : Production Part Approval Process. Used by automotive industry for acceptance of new products for relcase and use on automobiles.

PPCI (Peripheral Component Interconnect) : A standard interface used primarily on computer backplanes to connect interface cards and peripheral devices to the processor bus. PCI is often used for video display cards, network interfaces (e.g. Ethernet), and peripheral interfaces such as SCSI or USB. PCI buses typically also support the older Industry Standard Architecture (ISA) standard.

Pre-Bias Soft Start : A power-supply feature that prevents discharging of the output capacitor when the power supply starts up. Discharging the output capacitor could create either start up oscillation problems at cold start or large voltage disturbances on the output voltage bus at hot plug-in. Pre-bias soft start is an important feature in redundant power-supply systems, parallel power supply modules, battery back-up voltage buses, and other applications where multiple power sources supply one node.

Precipitation Hardening : Hardening and strengthening of a metal alloy by extremely small and uniformly dispersed particles that precipitate from a supersaturated solid solution.

Precipitation : The falling to earth of condensed water vapor in the form of rain, snow, sleet or hail.

Preemphasis : In some transmission and recording systems (e.g. vinyl records, FM radio, analog magnetic tape), there is more noise at higher frequencies. To offset this, the audio signal is "preemphasised" at the transmitter filtered with a high-pass filter to boost the higher audio frequencies. A matching low-pass filter is used at the receiver to return to an overall flat audio-frequency response. The filter at the receiver reduces the high-frequency noise introduced by the transmission process.

Pre-Ignition : Potentially damaging condition when ignition takes place ahead of the timed spark, often initiated by hot or incandescent deposits in the combustion chamber. The high temperatures induced by detonation can promote pre-ignition.

Prepreg : Continuous fiber reinforcement pre-impregnated with a polymer resin which is then partially cured.

Pressure Cooker Test : A Pressure Cooker Test (PCT) tests a part under high temperature, humidity, and pressure conditions. Also called an Autoclave Test or Pressure Pot Test (PPOT).

Pressure Sensor (MAP Sensor) : A sensor (transducer) producing an output signal according to inlet manifold pressure for use as a measurement of engine load by the ECU. May be contained within the ECU or may be a separate unit but in either case will have a pipe connection to the engine manifold. Term may also refer to a device providing the ECU with barometric pressure correction.

Primary Bond : Interatomic bonds that are relatively strong and for which bonding energies are relatively large.

Primary Coefficients : The factors relating distance to the basic electrical properties of transmission lines.

Primary Standards : Required drinking water quality standards related directly to human health. : These standards are required and enforceable by the U.S. : EPA.

Primary Treatment : Treatment which includes all operation prior to and including primary treatment, e.g., bar screening, grit removal, comminution, and primary sedimentation.

Printed Circuit Board : A printed circuit board, or PC board, or PCB, is a non-conductive material with conductive lines printed or etched. Electronic components are mounted on the board and the traces connect the components together to form a working circuit or assembly. A PC board can have conductors on one side or two sides and can be multi-layer a sandwich with many layers of conductors, each separated by insulating layers.

Procaryotic Organisms : Organisms which do not have a cellular membrane.

PROCHOT : Digital output pin on Intel's Pentium 4 processors that indicates the internal Thermal Control Circuit has activated. This occurs when the processor has reached its maximum safe operating temperature.

Producers : Autotrophic organisms which produce protoplasm using inorganic carbon and energy from the sun.

PROFIBUS : Vendor-independent open fieldbus standard used in manufacturing, building automation, and process control. Utilises a nonpowered two-wire (RS-485) network. PROFIBUS is standardised under the European Fieldbus Standard EN 50 170.

Programmable Delay Line : A delay line with a selectable amount of delay with out the need to change taps. The example below shows a passive delay line with a fixed input and output using three control lines to change the amount of delay via a multiplexer.

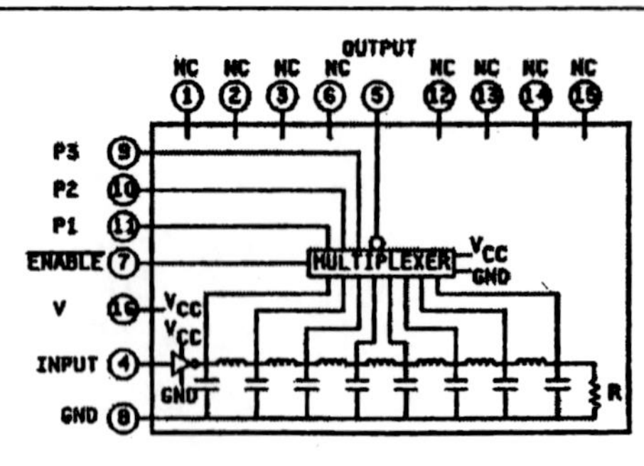

Programmable Delay Line

Propagation Coefficient : The complex factor which resolves into the attenuation and phase coefficients of a transmission line.

Propagation Delay : The time required for a signal to pass along a transmission line

Propagation Velocity : The velocity of wave propagation along a transmission line.

Proportional Limit : The point on a stress-strain curve at which the straight line proportionality between stress and strain seases.

P-type Semiconductor : A semiconductor for which the predominant charge carriers responsible for electrical conduction are holes. Normally, acceptor impurity atoms give rise to the excess holes.

Proximity Effect : The redistribution of current in a conductor brought about by the proximity of another conductor.

PTC (Positive Temperature Coefficient) : When the resistance of a component rises with temperature, it is said to have a positive temperature coefficient. Example: Hewlett-Packard's first commercial product, an audio oscillator, used a common light bulb as a PTC element in the feedback circuit to maintain constant output amplitude regardless of frequency.

Publicly Owned Treatment Works (POTW) : A municipal or domestic wastewater treatment facility.

Pulse Duration (Pulse Width) : The electrical pulse output produced by the ECU to open the injectors to admit a quantity of fuel according to the width or duration of the pulse. Pulse widths typically range from 2 or 3 milliseconds (thousandths of a second) at idle up to 8 or 9 milliseconds at full load.

Pulse-Amplitude Modulation : Pulse-Amplitude Modulation (PAM) is a pulse modulation technique in which the amplitude of the pulse is varied with the input signal amplitude.

Purge Valve : A valve which controls the purging action of evaporative fuel vapour from the evaporative loss charcoal canister.

Push-Button Switch **:** Style of switch that is actuated by depressing a button. The graphic shows a Panel Mount [chassis mount] Push-Button Switch. Styles of Push-Button switches include Momentary and Latching switches.

Push-Pull **:** An output structure which uses one active device to source current and a second device to sink current. Common examples are: a CMOS stage in which an n-channel device pulls toward ground or a negative supply and a p-channel device pushes current to bring the output up; an output stage in an audio amplifier with an NPN and PNP device in totem-pole configuration.

Push-Pull Amplifier **:** An amplifier that uses two transistors (or electron tubes) whose output signals are in phase opposition

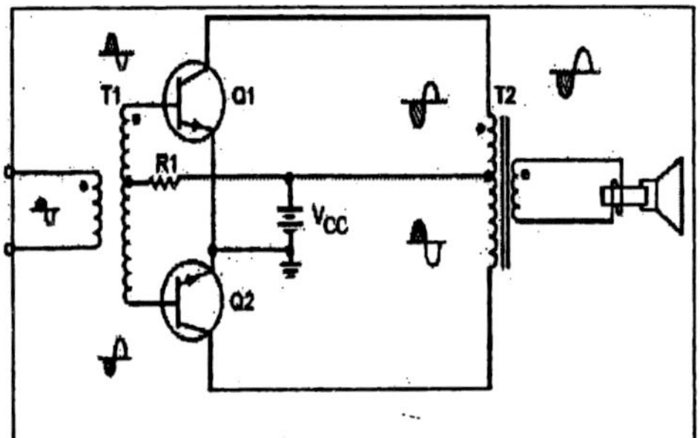

Push-pull Transistor Amplifier

PWM **:** (1) A method for using pulse width to encode or modulate a signal. The width of each pulse is a function of the amplitude of the signal. (2) A technique used to modulate the power delivered to a load. In DC-DC switching regulators, the pulse width driving the main power switch (and hence, the duty cycle) is varied to maintain the desired output voltage. In DC motor-control applications, pulse width is used to vary motor speed.

PWM Temperature Sensor **:** Temperature sensor with digital, logic-level output. The output has a fixed frequency and the duty cycle varies with the measured temperature.

Q

Q : Figure of merit of efficiency of a circuit or coil. Ratio of inductive reactance to resistance in servos. Relationship between stored energy (capacitance) and rate of dissipation in certain types of electric elements, structures, or materials.

QAM (Quadrature Amplitude Modulation) : A modulation method in which two signals are used to amplitude-modulate two carriers that are in quadrature (90 degrees out of phase with each other). The two modulated signals are combined. A common application is in PAL and NTSC color television transmission. Color is encoded into two analog signals (called I and Q), which modulate quadrature color carriers. Modems also use this approach, to increase the data bandwidth they can carry (or, more accurately, to trade bandwidth for error rate or noise immunity).

Q-Band : A band of frequencies in the range of 36GHz to 46GHz. Note that letter designations are no longer used.

Q-Meter : A device used to measure the Quality factor [Q] of a circuit. The Q-meter is not in wide spread use.

QPSK : *QPSK Symbol Constellation* Quadrature Phase Shift Keying (QPSK) is a form of Phase Shift Keying in which two bits are modulated at once, selecting one of four possible carrier phase shifts (0, 90, 180, or 270 degrees). QPSK allows the signal to carry twice as much information as ordinary PSK using the same bandwidth. QPSK is used for satellite transmission of MPEG2 video, cable modems, video-conferencing, cellular phone systems, and other forms of digital communication over an RF carrier.

QS-9000 : QS-9000 is the automotive quality standard surperseded by ISO/TS16949:2002 specification for component suppliers to the automotive industry.

Quad : A series of four separately insulated conductors, generally twisted together.

Quadrature Amplitude Modulation : Quadrature modulation in which the two carriers are amplitude modulated. (QAM):

Quadrature Phase Shift Keying (QPSK) : QPSK is a form of Phase Shift Keying in which two bits are modulated at once, selecting one of four possible carrier phase shifts (0, 90, 180, or 270 degrees). QPSK allows the signal to carry twice as much information as ordinary PSK using the same bandwidth. QPSK is used for satellite transmission of MPEG2 video, cable modems, videoconferencing, cellular phone systems, and other forms of digital communication over an RF carrier.

Quadrature : The state of being separated in phase by 90 degrees. Pertaining to the phase relationship between two periodic quantities varying with the same period, that is, with the same frequency or repetition rate, when the phase difference between them is one-quarter of their period.

Quality (of Sound) : The factor that distinguishes tones of pitch and loudness.

Quantisation : A process whereby the continuous range of input-signal values is divided into nonoverlapping subranges. Each of these subranges has a discrete value of the output uniquely assigned. Once a signal value falls within a given subrange, the output provides the corresponding discrete value.

Quantise : The process of restricting a variable to a number of discrete values.

Quantised Wave : A wave created by the arbitrary division of the entire range of amplitude (or frequency, or phase) values of an analog wave into a series of standard values. Each sample takes the standard value nearest its actual value when modulated.

Quantising Noise : Distortion introduced by quantising the signal.

Quantum-Mechanical Tunneling : The action of an electron crossing a PN junction because of tunnel effect.

Quartz Oscillator : An oscillator in which a quartz crystal is used to stabilise the frequency.

Quenching : The cooling of a piece of steel in water or oil from a certain temperature to harden or temper it.

QuERC : Software that examines bias and transient simulation output and flags devices operating above limits.

Quick Connect Terminal : A terminal or receptacle used for rapid inter-connection. Normally a flat metal blade is used as a quick connect terminal, the receptacle slides over metal blade using pressure to make the connection.

Quiescence : The state of an amplifier with no signal applied. The operating conditions that exist in a circuit when no input signal is applied to the circuit.

Quiescent : For an electronic circuit, a quiet state in which the circuit is driving no load and its inputs are not cycling. Most commonly used for the specification "quiescent current," the current consumed by a circuit when it in a quiescent state.

Quiescent Current : The current flowing in a circuit during a period in which a circuit element is not performing an active function in the system.

Quiescent State : The period when a transistor, tube, or other circuit element is not performing an active function in the circuit.

R

R-2R : (1) Short for R-2R ladder: A method for D/A conversion which employs a ladder-shaped resistor array composed of two resistor values: R and 2R. Each bit in the digital input switches a ladder's rungs in and out of the network to change the output voltage by an amount proportional to the significance of the bit. (2) Rail-to-rail.

R2r Ladder : A resistor network made by individual resistors or a resistor package.

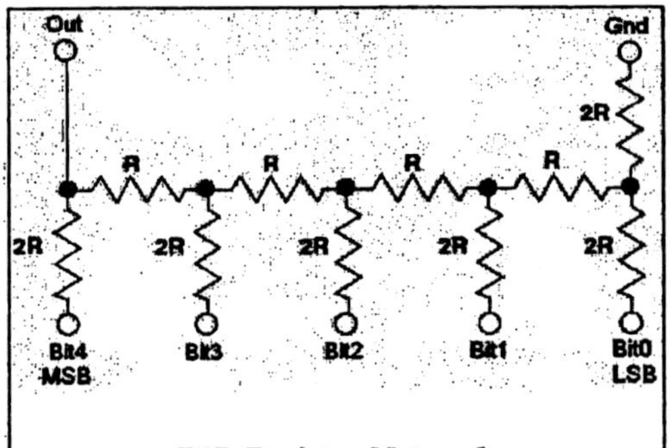

R2R Resistor Network

Race Condition : defines a condition when a device's output depends on two or more nearly simultaneous events to occur at the input.

Raceway : A channel designed expressly for holding wires, cables, or busbars, with additional functions as permitted in this subpart. Raceways may be of metal or insulating material, and the term includes rigid metal conduit, rigid nonmetallic conduit, intermediate metal conduit, liquid-tight flexible metal conduit, flexible metallic tubing, flexible metal conduit, electrical metallic tubing, under-floor raceways, cellular concrete floor raceways, cellular metal floor raceways, surface raceways, wireways, and busways.

Rack **:** A vertical frame that holds or supports equipment. If the rack uses side panels, than it's called a rack enclosure.

Radar **:** An acronym for RAdio Detecting And Ranging.

Radial Engine **:** Cylinders arranged radially around the crankshaft. In this case a 1850hp Wright R-2600-29 Cyclone 14-cylinder radial. Note that this is effectively two 7 cylinder radials back to back with the rear bank of cylinders being rotated half the cylinder to cylinder spacing.

Radiation Field **:** The electromagnetic field that radiates from an antenna and travels through space.

Radiation Losses **:** The losses that occur when magnetic lines of force about a conductor are projected into space as radiation and are not returned to the conductor as the cycle alternates.

Radiation Resistance **:** The resistance that if inserted in place of the antenna would consume the same amount of power as that radiated by the antenna.

Radiation **:** This is the electromagnetic radiation emitted by a body as a function of its temperature. The idealised body is a black body, this being one which absorbs all radiation that impinges on it. Stefan's law states that the total energy radiated per unit time per unit surface area of a black body is proportional to the fourth power of the temperature (kelvins) of the body. E=sT4. where = Stefan's constant = 5.67x10-8Wm-2K-4.

Radio Communications **:** The term describing teletypewriter, voice, telegraphic, and facsimile communications.

Radio Frequency (RF) **:** Any frequency of electromagnetic energy capable of propagation into space. The frequencies that fall between 3 kilohertz and 300 gigahertz used for radio communications.

Radio Frequency Carrier Shift : The system that uses a keyer to shift a radio frequency signal above or below an assigned frequency. These shifts correspond to marks and spaces.

Radio Horizon : The boundary beyond the natural horizon in which radio waves cannot be propagated over the earth's surface.

Radio Set Control Unit : Equipment used to remotely control certain transmitter and receiver functions.

Radio Waves : A form of radiant energy that can neither be seen nor felt. An electromagnetic wave that is generated by a transmitter.

Radix Point : Also called binary point, octal point, decimal point, and so forth, depending on the number system.

Radix : Also called the base. The number of distinct symbols used in a number system. For example, since the decimal number system uses ten symbols (0, 1, 2, 3, 4, 5, 6, 7, 8, 9), the radix is 10. In the binary number system, the radix is 2 because it uses only two symbols (0, 1).

RAID (Redundant Array of Independent Disks) : A redundant array of inexpensive disks. RAID is a performance-enhancing method of storing the same data in different places on multiple hard disks to achieve speed and/or data redundancy.

Rail-to-Rail Input : The allowable input signal range includes the supply voltages.

Rail-to-Rail Input or Output : The allowable input and output voltage ranges include the power-supply rails.

Ram Induction : A technique for harnessing resonant pulses created by the induction process to improve volumetric efficiency, dependent on the length, cross section, and sometimes the interconnection, of passageways. A good ram system can boost torque by as much as 20%. Not to be confused with a forward facing air intake.

Random Copolymer : A polymer in which two different mer units are randomly distributed along the molecular chain.

Random Jitter : Random jitter (RJ) includes all jitter components not defined as deterministic jitter.

Raw Emissions : Emissions direct from the engine before treatment by a catalyst or similar.

RC : Resistance-capacitance; resistor-capacitor. In particular, an RC network is a network composed of resistors and capacitors in a series-parallel combination, usually to filter or delay a signal.

Reactive Waste : A waste which; (1) reacts violently with water, (2) forms potentially explosive mixtures with water, (3) is normally unstable, (4) contains cyanide or sulfide in sufficient quantity to evolve toxic fumes at high or low pH, (5) is capable of exploding if heated under pressure, or (6) is an explosive compound listed inDepartment of Transportation (DoT) regulations. One of EPA's four hazardous waste properties.

Reaeration : The dissolving of molecular oxygen from the atmosphere into the water.

Receiver : A circuit that accepts signals from a transmission medium and decodes or translates them into a form that can drive local circuits. Examples: A radio receiver that detects and demodulates the signal from the airwaves. An ultrasonic receiver that turns ultrasound signals into electrical signals.

Receiving Water Quality Standards : Standards which require a discharger to maintain a certain quality level in the receiving water.

Receiving Water : A water which receives wastewater (treated or otherwise) discharges.

Recovery Time : The time for a sensor to return to baseline value after the step removal of the measured variable. Usually specified as time to fall to 10% of final value after step removal of measured variable.

Recrystallisation : The formation of a new set of strain-free grains within a previously cold-worked material; normally an annealing heat treatment is necessary.

Recycling : The recovery and reuse of a product which would otherwise be thrown away.

REF : REF is a term that appears on IC package drawings in reference to dimensions. It stands for REFERENCE and indicates that this is a reference dimension, calculated or based on another dimension. For example, the dimension from the first pin to the last pin in the row of a DIP (dual inline package) usually is tagged as REF because it is a multiple of the distance from pin-to-pin. In the case of a 16-pin DIP, the first pin to last pin dimension is 7 times the pin-to-pin dimension (7 spaces between 8 pins).

Reflection : Deflection of a light beam at the interface between two media.

Reformulated Gasoline : A new 'low emissions' type of unleaded gasoline fuel introduced by Atlantic Richfield Oil in the US in 1989 with the ability to reduce emissions from older vehicles without any emission control systems. It had low volatility to reduce evaporative emissions, low benzene and aromatic content to reduce toxic emissions, and contained oxygenates to reduce HC and CO emissions. Now widely used and mandatory in some parts of the world.

Refraction : Bending of a light beam upon passing from one medium into another.

Refractive Index : The ratio of the velocity of light in a vacuum to the velocity of light in some medium.

Refractory : A metal or ceramic that may be exposed to extremely high temperatures without deteriorating rapidly or without melting.

Refuse Derived Fuel (RDF) : A fuel derived from the combustible portion of municipal solid waste. The fuel is often processed into small briquettes, similar in size to charcoal.

Regenerative Detector : A detector circuit that produces its own oscillations, heterodynes them with an incoming signal, and deflects them.

Regenerative Feedback : The process by which a portion of the output signal of an amplifying device is fed back in phase to reinforce the input.

Regulator : The section in a power supply that maintains the output of the power supply at a constant level in spite of large changes in load current or input line voltage. There are two main styles; Switching Regulators and Linear Regulators.

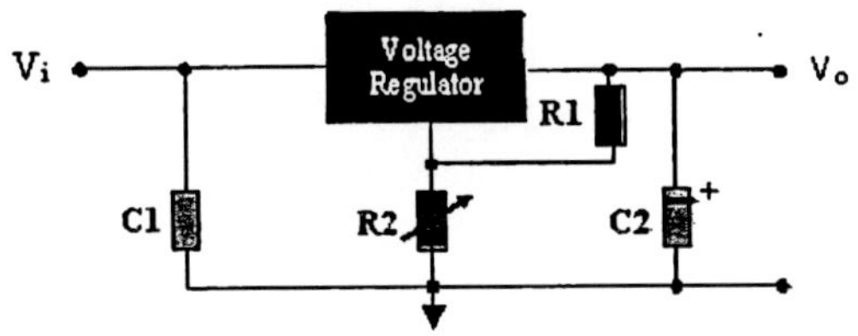

Voltage Regulator

Relative Magnetic Permeability : The ratio of the magnetic permeability of some medium to that of a vacuum.

Relay : A relay is an electromagnetic switching device consisting of an armature which is moved by an electromagnet to operate one or more switch contacts. Some advantages of relays are that they provide amplification and isolation and are straightforward. They can switch difficult voltages (e.g. RF or high-powered AC) with complete isolation and no worries about level translation.

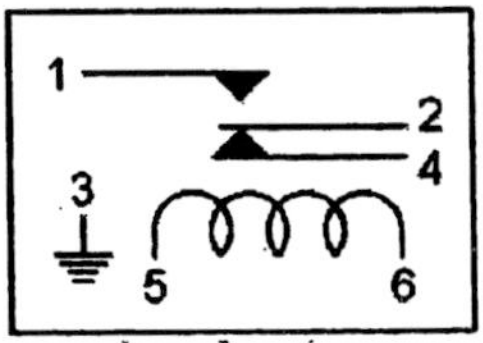

Relay Schematic

Remote Diode : A diode or diode-connected bipolar transistor used as a temperature-sensing element, often integrated onto an integrated circuit whose temperature is to be measured.

Remote Temperature : Temperature at a location other than at the die of the temperature-measuring integrated circuit.

Remote Temperature Sensor : A remotely located PN junction used as a temperature sensing device, usually located on an integrated circuit other than the one doing the measurement.

Repeater : A station that receives a signal [on one frequency] and re-transmits the signal [on another frequency] over the air. A device that regenerates the incoming signal to the original signal level, and than retransmits that signal to increase the distance the signal can travel.

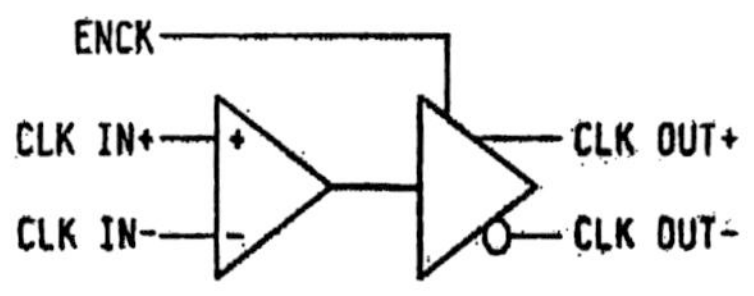

Differential Repeater

Resistance : Resistance, represented by the symbol R and measured in ohms, is a measure of the opposition to electrical flow in DC systems. Resistance is the voltage across an element divided by the current (R = V/I).

Resistance Unbalance : A measure of the inequality of the resistance of the two conductors of a transmission line.

Resistivity : The reciprocal of electrical conductivity, and a measure of a material's resistance to the passage of electric current.

Resource Smart : ResourceSmart is a design approach for green engineering and smarter use of resources. ResourceSmart is not just about lower-power components, it's about all the resources an engineer manages: energy, materials, and the human creative energy that drives new products. No one is more ResourceSmart than an engineer. ResourceSmart strategy combines tutorials, application notes, and technical articles with resource-friendly products and techniques.

Respiration **:** Energy production in which oxygen is the terminal electron acceptor, i.e. oxidation to produce energy where oxygen is the oxidising agent.

Response Time **:** The time for a sensor to respond from no load to a step change in load. Usually specified as time to rise to 90% of final value, measured from onset of step input change in measured variable.

Return Loss **:** The difference between the power incident upon a discontinuity in a transmission system and the power reflected from the discontinuity.

Reverse Recovery Time **:** When switching from the conducting to the blocking state, a diode or rectifier has stored charge that must first be discharged before the diode blocks reverse current. This discharge takes a finite amount of time known as the Reverse Recovery Time, or trr. During this time, diode current may flow in the reverse direction.

Reversible Reaction **:** A reaction in which the reactant(s) proceed to product(s), but the products react at an appreciable rate to reform reactant(s).

RF (Radio Frequency) **:** An AC signal of high enough frequency to be used for wireless communications.

RF (Radio Frequency) Amplifier **:** An amplifier designed to amplify signals with frequencies between 10 kilohertz (10 kHz) and 100,000 megahertz (100,000 MHz).

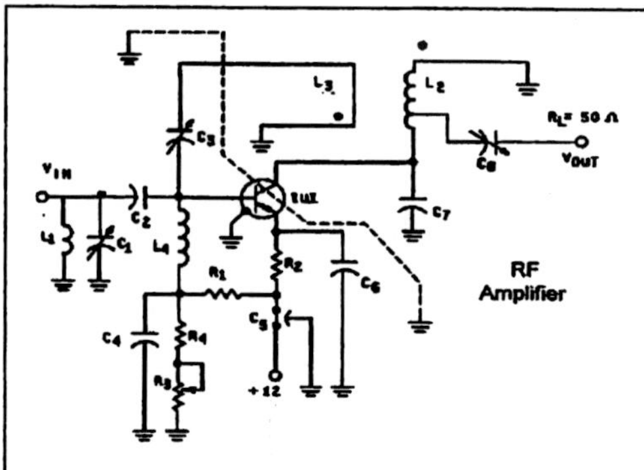

RF Amplifier

RFID (Radio Frequency Identification) : A method for uniquely identifying an object using a tag or module that carries a unique ID number, or code. Identification can be made using wireless (RF, or radio-wave) connection, meaning no line-of-sight or physical contact is needed. There are many different ways to achieve RFID and many applications including pet ID, identification of parts on an assembly line, tracking goods in manufacturing or retail settings, etc.

Ripple Rejection : Ripple Rejection is the ability of an amplifier to maintain accurate output voltage despite AC fluctuations in the power supply.

Ripple Voltage : The alternating component of unidirectional voltage. This component is small compared to the direct component.

RISC (Reduced Instruction Set Computer) : Computer hardware designed to support a short list of simple instructions. This makes the hardware simpler and faster, since it does not need to accommodate complex instructions. Although more instructions must be executed for some operations, a RISC architecture can be faster, depending on the instruction mix, the design of the instruction set, and how effective the compilers and support software are in translating operations into optimised instructions.

Rotary Engine : Rotary engines were used extensively in early aircraft. Note that with a rotary engine the crankshaft is fixed and does not rotate, the propeller is fixed to the crankcase and the whole crankcase rotates with the propeller.

RSSI (Received Signal Strength Indicator (or Indication)) : A signal or circuit that indicates the strength of the incoming (received) signal in a receiver. (The signal strength indicator on a cell phone display is a common example). RSSI is often done in the IF stage before the IF amplifier. In zero-IF systems, it is done in the base band signal chain, before the base band amplifier. RSSI output is often a DC analog level. It can also be sampled by an internal ADC and the resulting codes available directly or via peripheral or internal processor bus.

RTCs (Real-time clock) **:** Integrated circuit that contains a timer that supplies the time of day (and often, the date). An RTC generally contains a long-life battery to allow it to keep track of the time even when there is no power applied.

RTD **:** A Resistance Temperature Detector (RTD) is a device with a significant temperature coefficient (that is, its resistance varies with temperature). It is used as a temperature measurement device, usually by passing a low-level current through it and measuring the voltage drop. A thermistor is a common type of RTD.

Runoff **:** The water that flows overland to lakes or streams during and shortly after a precipitation event.

Rupture **:** Failure that is accompanied by significant plastic deformation.

RZ (Return to Zero) **:** A binary bitstream encoding scheme in which the signal returns to zero voltage in between the data bits. The signal has three valid levels: High, Low, and the return to zero volts after each bit.

S

S : 1. Siemen(s), standard unit for conductance 2. Lower case is the standard abbreviation for seconds.

Saltwater Intrusion : The gradual replacement of freshwater by saltwater in coastal areas where excessive pumping of groundwater occurs.

Sample Rate : The rate at which a signal is sampled. The frequency at which an analog signal is sampled, as in a Digital-to-Analog Converter.

Samples per Second : (1) sps: Samples per second. In data conversion, an analog signal is converted to a stream of numbers, each representing the analog signal's amplitude at a moment in time. Each number is called a "sample." The number sample per second is called the sampling rate, measured in samples per second. (2) ksps: Kilosample(s) per second (thousands of samples per second) (3) Msps: Megasamples per second (millions of samples per second).

Sampling Oscilloscape : An oscilloscope that samples the measured signal and reconstructs it as a visual representation. All Oscilloscopes produced for the last few decades are Sampling Scopes.

Sampling Rate : An A/D converter converts an analog signal into a stream of digital numbers, each representing the analog signal's amplitude at a moment in time. Each number is called a "sample." The number sample per second is called the sampling rate, measured in samples per second.

Sampling Rate **:** An A/D converter converts an analog signal into a stream of digital numbers, each representing the analog signal's amplitude at a moment in time. Each number is called a "sample." The number sample per second is called the sampling rate, measured in samples per second.

SAN (Storage Area Network) **:** A network infrastructure of shared multihost storage, linking all storage devices and interconnecting remote sites.

SAR (Successive Approximation Register) **:** Used to perform the analog-to-digital conversion in successive steps in many analog-to-digital (ADC) converters.

Satellite Eclipse **:** An eclipse where the rays of the sun do not reach the satellite. This prevents recharging of the solar cells of the satellite and decreases the power to the transmitter.

Satellite **:** A radio relay station that orbits the earth. A complete satellite communications system also includes earth stations that communicate with each other via the satellite. The satellite receives a signal transmitted by an originating earth station and retransmits that signal to the destination earth station(s). Satellites are used to transmit telephone, television and data signals originated by common carriers, broadcasters and distributors of cable TV program material.

Satellite-Sun Conjunction **:** A period when the satellite and sun are close together and the noise from the sun prevents or hampers communications.

Saturable-Core Reactor **:** A coil in which the reactance is controlled by changing the permeability of the core.

Saturated Steam **:** Steam at the same temperature as the boiling point of water at that particular pressure is said to be saturated. Scale : The debate as to what is scale and what is not will go on and on. We have generated a list of different scales.

Saturation **:** The condition existing in any circuit in which an increase in the input signal produces no further change in the output. The operating point of a vacuum tube or transistor at which a further increase in grid or base current no longer produces an increase in plate or collector current. In a magnetic core, the condition in which a magnetic material has reached a maximum flux density and the permeability has decreased to a value of approximately 1.

SAW (Surface Acoustic Wave) : A sound wave that propagates along the surface of a solid and is contained within the solid. SAW devices typically combine compressional and shear components. In Wireless applications, SAW refers to a Surface Acoustic Wave band-pass filter, which exhibits much better out-of-band rejection, but has higher passband ripple and insertion loss.

Saw Tooth : A waveform that appear like teeth on a saw. A Saw-Tooth waveform has a linear rise time and linear fall time.

Scaling Factor : The term used to describe the use of unequal resistors in a servo's summing network to compensate for differences between input and output signal levels.

Scan Design : A design technique in which the internal registers or flip-flops of a circuit can be chained, to allow an external circuit to easily read and write their contents. When internal memory elements are not directly accessible from the circuit's outside pins, testing is difficult because their state is unknown. With scan design, a signal reconfigures the elements into a "scan chain" and their contents can be read and if desired, altered.

SCART : Also known as Euroconnector or Peritel, a 21-pin connector commonly used in Europe to interconnect satellite receivers, television sets, and other audiovisual equipment (e.g. videocassette recorders). A single connector combines audio and video signals.

Schematic Symbols : A letter, abbreviation, or design used to represent specific characteristics or components on a schematic diagram.

Schematic : A diagram which shows, by means of graphic symbols, the electrical connections and functions of a specific circuit arrangement.

Schmitt Trigger : An integrated circuit which has hystersis between an upper voltage limit and lower voltage limit and does not switch until one of those limits are reached.

Schottky Diode : A diode realised via a "Schottky-barrier junction" a metal-semiconductor junction rather than the P-N junction used by conventional semiconductor diodes. Schottky diodes are often chosen for their high switching speed and low forward voltage drop.

Screen Grid : A grid placed between a control grid and the plate and usually maintained at a fixed positive potential. Used with Vacuum tubes or CRTs.

Screening Attenuation : The decibel difference between the signal level present at the input of a device, and the radiated power originating from the device.

SCSI : Small Computer System Interface (pronounced "scuzzy"), an interface standard for connecting peripheral devices to computers. Hardware components for implementing a SCSI interface include connector ports on computers and cables for connecting peripheral devices to the computer. SCSI is gradually being supplanted by the newer USB and IEEE-1341 standards.

SCT (Single Chip Transceivers) : A single IC that includes data communication transmitter and receiver functions.

SD : (1) Signal detect: An output that indicates when a signal is present. A form of Signal Strength Indicator. (2) Secure Digital, a media format for nonvolatile external memory. The successor to the "MultiMedia Card" format, or MMC, SD card memories typically operate from 3.3V supplies with modest current requirements. SD memory cards are best known as storage for digital cameras, smart phones, and other consumer electronic devices.

SDTV (Standard Definition Television) : Digital formats that do not achieve the video quality of HDTV, but are at least equal, or superior to, NTSC pictures. SDTV may have either 4:3 or 16:9 aspect ratios, and includes surround sound. Variations of fps (frames per second), lines of resolution, and other factors of 480p and 480i make up the 12 SDTV formats in the ATSC standard.

Second Detector (Demodulator) : The part of the receiver that separates the audio or video component from the modulated intermediate frequency.

Second Harmonic Distortion : ***Second Harmonic Distortion (HD2)***: Ratio of second-order harmonic to the input signal (carrier). Often measured as dBc.

Secondary Cell : A cell that can be recharged by a current being passed through the cell in a direction opposite to the discharge current.

Secondary Coefficients : Attenuation and phase coefficients and characteristic impedance of a transmission line.

Secondary Emission : The liberation of electrons from an element, other than the cathode, as a result of being struck by other high-velocity electrons.

Secondary Loop : In a cooling system, the loop that transfers the heat from the heat source, such as electronic equipment, to the primary loop; usually distilled water.

Secondary Standards : Recommended drinking water quality standards which relate to aesthetics and/or health. These standards are recommended, not required.

Secondary Treatment : In wastewater treatment, the conversion of the suspended, colloidal and dissolved organics remaining after primary treatment into a microbial mass with is then removed in a second sedimentation process. Secondary treatment included both the biological process and the associated sedimentation process.

Secured Landfill : A landfill which has containment measures such as liners and a leachate collection system so that materials placed in the landfill will not migrate into the surrounding soil, air and water.

Sedimentation : The gravity settling, and thus removal, of materials more dense than the suspending fluid.

Self Diagnosis : The ability of advanced systems to record certain failure modes which can then be displayed, usually in a simple code or as a fault number, to assist diagnosis.

Semiconductor : 1. A substance that can act as an electrical conductor or insulator depending on chemical alterations or external conditions. Examples are silicon, germanium, and gallium arsenide. Also called "III-V" materials since semiconductor elements are in groups III and V of the periodic table of chemical elements. 2. An electronic device (e.g. a transistor, diode, or integrated circuit) manufactured from semiconductor materials. Semiconductor devices control and amplify because a small voltage or current, or a physical stimulus (such as light or pressure), allows the semiconductor to pass or block electrical current. Devices can be fabricated with other capabilities such as passing electric current in only one direction, emitting light, mixing and transforming signals, etc.

Sense Resistor : A resistor placed in a current path to allow the current to be measured. The voltage across the sense resistor is proportional to the current that is being measured and an amplifier produces a voltage or current that drives the measurement.

SEPIC (Single Ended Primary Inductor Converter) : A DC-DC converter topology that acts both as a boost and a buck converter (that is, will step up or down, depending on the input voltage).

Sequential Injection : System in which each injector is fired independently at the optimum point for its particular cylinder. Requires separate wiring to each injector. Has advantages for emissions and drivability in certain circumstances.

Serial Interface : A serial interface (as distinguished from a parallel interface) is one in which data is sent in a single stream of bits, usually on a single wire-plus-ground, wire-pair, or single wireless channel (or two sets, one for each direction). Examples include USB, RS-232, I2C, and 1-Wire. By contrast, a parallel interface sends several bits at once, on separate wires.

Serial Link : A means of communication between computers and accessories in which data is transferred as a series of timed pulses along one wire in a connection conforming to a recognised standard such as RS232. A common method is for each byte (8 bits) of data to be preceded and followed by register pulses to maintain synchronisation. Serial data transfer, although slower than the parallel system within the computer, is much more reliable and convenient over any distance. A serial link is widely used for connecting interrogation instruments to vehicles for diagnostic or tune up purposes. Such a link can be made with just three wires, one for data in, one for data out, and ground.

SFDR (Spurious-Free Dynamic Range) : A term used to specify A/D and D/A converters (ADCs and DACs). In ADCs, Spurious-Free Dynamic Range (SFDR) is the ratio of the RMS amplitude of the carrier frequency (maximum signal component) to the RMS value of the next largest noise or harmonic distortion component. SFDR is usually measured in dBc (with respect to the carrier frequency amplitude) or in dBFS (with respect to the ADC's full-scale range). I

SHA (Secure Hash Algorithm) : A message digest algorithm developed by the NSA for use in the Digital Signature standard, FIPS number 186 from NIST. SHA is an improved variant of MD4 producing a 160-bit hash. SHA is one of two message digest algorithms available in IPSEC.

Sheet Metal : The traditional way of defining the thickness of sheet metal has been the *Birmingham Sheet Metal Gauge.*

Shift Register : Two or more bistable elements (flip-flops) connected in series. With each tick of the clock, the output of stage n is shifted to stage n+1. Applications include clock or signal delays, delay lines, linear-feedback shift registers.

Shock Load : Influent wastewater entering the plant which has an unusually high organic content and/or high flow rate.

Shock Sensor : An acceleration sensor, generally a piezoelectric type, that can measure high acceleration but cannot measure static g forces.

Shoot-Through Current : In a push-pull amplifier stage, one transistor pushes current to the output to drive it toward a positive voltage; a second device pulls down. These are designed so both devices are never fully on, which would effectively short the power supply. The rush of current that occurs while both devices are on is called the shoot-through current. Events that allow both devices to be on (e.g. circuit faults or a brief moment in the switching cycle) are said to "crowbar" the circuit because of its similarity to a power supply protection circuit.

Shunt : A small plastic female Jumper [picture to right]. A shunt is normally a two terminal component on 0.1 inch centers which is fitted over a dual-row headed with the same dimensions.

Shunt Voltage Regulator : A regulator whose regulating device is in parallel with the load resistance. 3-Terminal Shunt Regulator shown.

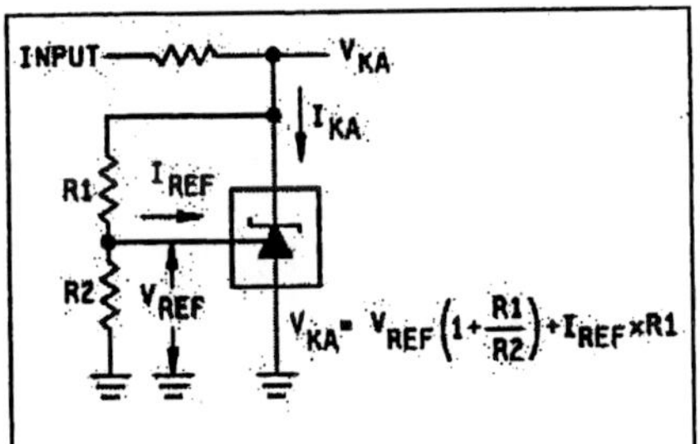

3-Terminal Shunt Regulator

Shutdown : A feature of many ICs, typically controlled via a logic-level input, which dramatically reduces power consumption when the device is not in use.

Sigma Delta : An ADC architecture consisting of a 1-bit ADC and filtering circuitry which over-samples the input signal and performs noise-shaping to achieve a high-resolution digital output. The architecture is relatively inexpensive compared to other ADC architectures.

Signal Conditioner **:** A device placed between a signal source and a readout instrument to change the signal. Examples are attenuators, preamplifiers, charge amplifiers, and sophisticated level-translating deivices that can compensate for non-linearities in the sensor or amplifier.

Signal-Invalid O/P **:** Signal invalid output. Indicates when all RS-232 signals to the IC are in the invalid range.

Signal-to-Noise Ratio **:** Signal-to-Noise Ratio, the ratio of the amplitude of the desired signal to the amplitude of noise signals at a given point in time. The larger the number, the better. Usually expressed in dB.

Silicon **:** After carbon silicon is the most abundant element on earth. Silicon is obtained by reducing sand (silica: SiO_2), with carbon.

Silicon-controlled Rectifier (SCR) **:** A semiconductor device that functions as an electrically controlled switch. A 3-terminal semiconductor device consisting of a Gate, Anode and, Cathode.

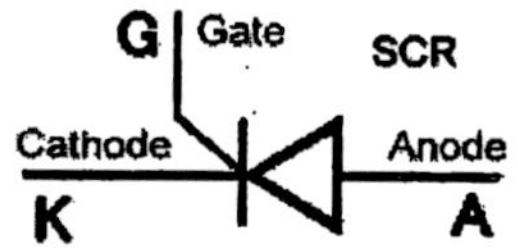

3-terminal Semiconductor Device

Silver Steel **:** The word "silver" in the name comes from the appearance of this high carbon steel manufactured to BS 1407.

SINAD (Signal-to-noise and distortion ratio) **:** The RMS value of the sine wave f(IN) (input sine wave for an ADC, reconstructed output sine wave for a DAC) to the RMS value of the converter noise from DC to the Nyquist frequency, including harmonic content. It is typically expressed in dB (decibels).

Single Point or Throttle Body Injection : System where injection takes place at the throttle body only and mixture is distributed to the individual cylinders by the inlet manifold, resembling a carburetter system.

Site Remediation : The process of cleaning up a hazardous waste disposal site that has either been abandoned or that those responsible either refuse to cleanup or are financially unable to cleanup.

Siting : Obtaining government (federal, state, and local) permission to construct an environmental processing, treatment, or disposal facility at a given site.

Skew : A measure of the inequality of propagation delay between two pairs in a multi-pair cable.

Skin Effect : The tendency of alternating currents to increasingly flow nearer the surface of a conductor as frequency increases.

Sludge : A mixture of solid waste material and water.Sludges result from the concentration of contaminants in water and wastewater treatment processes. Typical wastewater sludges contain from 0.5 to 10 percent solid matter. Typical water treatment sludges contain 8 to 10 percent solids.

Smart Battery : A battery with internal circuitry that provides level of charge status to the host system.

Smart Phone : A phone with a microprocessor, memory, screen, and built-in modem. The smart phone combines some of the capabilities of a PC in a handset device and typically include Internet connectivity.

Smart Signal Conditioner : Signal conditioner that is programmable or has a flexible architecture to allow it to accomplish sophisticated signal transformations and corrections.

SMD : (1) Surface Mount Device (SMD): An electronic component that mounts on the surface of a printed circuit board (as opposed to "through-hole" components which have pins that are inserted into holes). SMDs typically allow more components per square centimeter of PC board, but their scale is such that hand assembly and prototyping may be difficult. (2) Standard Military Drawing (SMD): A U.S. government program for standardised MIL-STD-883 product specifications, to simplify military procurement. Sponsored by the DSCC (Defense Supply Center, Columbus).

SMR (Specialised Mobile Radio) : Indicates the 896MHz to 901MHz band (800MHz band), which uses two paired 25kHz channels, and the 935MHz to 940MHz band (900 MHz band), which uses two paired 12.5kHz channels. Ten 20-channel blocks have been allocated in these frequency bands by the FCC. 900MHz SMR is primarily used for radio dispatch, paging, and wireless data communications.

SOHO (Small Office/Home Office) : Businesses that are either run from home or a from a small office. Software and hardware companies sometimes promote products as suitable for the SOHO market.

Solid State : A solid state device or circuit is one that relies on semiconductors rather than mechanical or vacuum tube circuits.

SONET (Synchronous Optical Network) : A North American standard for transmission in synchronous optical networks. It defines a family of rates, formats, interfaces, transport options, and maintenance capabilities. The minimum rate for SDH is 155Mbps.

Source : (1) One of the three terminals that comprise a FET. A voltage on the gate controls the current flow between the source and drain. (2) A "voltage source" or "current source" is an idealised generator of voltage or current in a circuit. These sources can be AC or DC, controlled or constant.

Source Reduction : The elimination or reduction of the waste at the source by modification of the actual process which produces the waste.

Space Diversity : In radio systems, Space Diversity transmits a signal on multiple propagation paths.

S-Parameters : The reflection and transmission coefficients used in impedance matching between high-speed (RF) devices and transmission lines/traces.

Speaker: A device that radiates acoustic energy into the air.

Speaker System

Specific Heat Capacity : The amount of heat required to raise unit mass of a substance by one degree.

Specific Volume : The volume occupied by a unit mass of a substance. This is the reciprocal of density and the units as such would be m3/kg.

Spectrum Analysis : The display of electromagnetic energy arranged according to wavelength or frequency.

Spectrum Analyzer : A test instrument that provides a visual display of the frequency distribution of an RF signal such as a transmitter output.

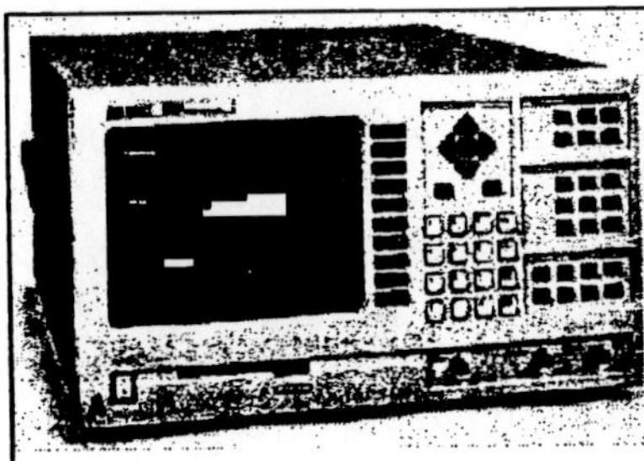

Spectrum Analyzer

Speed Law : The way in which fueling or ignition advance changes according to engine speed, often influenced by harmonic effects in the induction and exhaust processes or by the need to control exhaust emissions.

Spread Spectrum : A technology that modulates a signal over many carrier frequencies at once. This method can be used to make transmissions more secure, reduce interference, and improve bandwidth-sharing. Spread-spectrum techniques can also be used to reduce electromagnetic interference by dithering the clock frequency so emissions are no longer concentrated at one frequency.

SQC (Statistical Quality Control) : Use of statistical methods to measure and improve the quality of manufacturing processes and products. The term "statistical process control" is often used interchangably.

Square Wave : A rectangular shaped periodic waveform which switches between to fixed values.

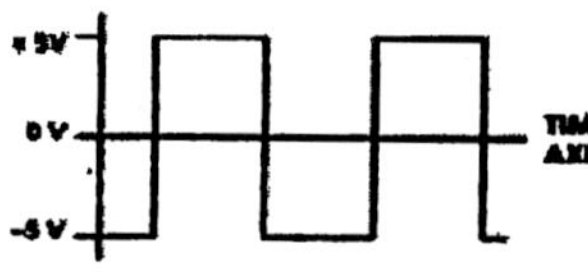

Square Wave

Squelch : A circuit which mutes the signal when it is below a certain level. Typically used to quiet the signal when only noise is present.

Squish : Turbulence induced into the air/fuel charge within the combustion chamber as the piston comes close to some part of the combustion chamber surface as it approaches TDC. Any gas contained in this zone (squish zone) is squeezed out into the main combustion chamber and promotes vigorous charge movement therein. Mainly useful for the later stages of combustion.

Stainless Steel : A range of stainless steels are available that have varying degrees of resistance to corrosion, containing between 8 and 25% Chromium. Other elements used in Stainless Steel are Nickel and Niobium.

Star Ground : A pcb layout technique in which all components connect to ground at a single point. The traces make in a "star" pattern, emanating from the central ground.

STB : A "set top box," or STB, is a generic name for an electronic interface between a cable television or satellite signal and video display and recording devices. Typically a box that can be placed atop the television set (hence the name), it can have many functions, including acting as a tuner, decoding digital or analog television signals, removing encryption, and allowing the purchase of pay-per-view channels.

STC : (1) Silicon Timed Circuit: A circuit that produces a delayed version of the input signal. Also known as a delay line. (2) System Timing and Control: Clock generation and distribution systems and components. May include the means for clock control such as spread-spectrum clock generation for EMI reduction, skew rate control, rate dividers, rate control, width, delay, and phase adjustment.

Stereo Jack : A jack designed to handles two channels. There are many different sizes and styles of stereo jacks, one example.

1/4 inch Stereo Phono Jack

Stepper Motor : A type of electric motor having multiple windings energised by pulses causing the armature to rotate in small increments. Often used to control idle air supply.

Step-Up DC-DC : A switch-mode voltage regulator in which output voltage is higher than its input voltage.

Stirling Engine : An external combustion engine. Also known as a Hot Air Engine. The Stirling engine was invented in 1816 by a Scotch clergyman, Robert Stirling, as a competing technology to the steam engine. Stirling engines work by subsequently heating and cooling a gas inside a closed volume. As the gas is heated it expands and as it is cooled it contracts. This expansion and contraction works forces a working piston which in turn drives a crankshaft.

Stoichiometric. Term used to describe a chemically correct mixture containing excess of neither air nor fuel (Lambda = 1). Stoichiometric fueling provides the ideal conditions for operation of a three-way catalyst, the fueling being controlled by Lambda feedback. Otherwise the condition has little merit because best economy from gasoline is obtained with fueling that is weaker by about 15% and best power is obtained when about 15% richer.

Stratified Charge Combustion : A type of Lean Burn combustion system in which the air/fuel ratio is deliberately divided into a relatively rich zone in which combustion can easily become established, with the extremities containing much weaker mixture. The overall air/fuel ratio can therefore be considerably weaker than would normally be possible. Stratified combustion can be achieved by careful control of swirl and fuel admission or by using a divided combustion chamber with the richer mixture confined to a pre-chamber containing the spark plug.

Stratosphere : The atmosphere from approximately 12 km to 70 km. The temperature of the atmosphere increases in this region.

Strong Acid : An acid that, for practical purposes, ionises completely under the conditions of interest. : Common strong acids are hydrochloric, nitric, and sulfuric.

Structural Return Loss : A measure of the uniformity of a transmission line's impedance.

Substrate Level phosphorylation : The synthesis of the energy storage compound adenosine triphosphate (ATP) from adenosine diphosphate (ADP) using organic substrates without molecular oxygen.

Summing Network **: A combination of two or more parallel resistors used in servosystems as an error detector. The output of the network is the algebraic sum of the inputs. The circuit example shows an Op Amp summing circuit.**

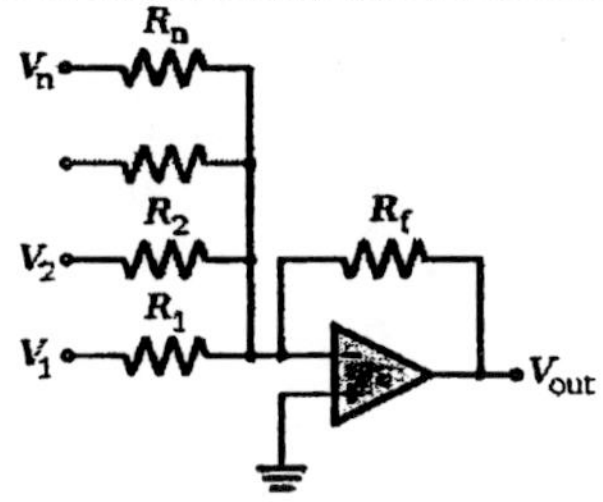

Op-Amp Summing Circuit

Superheated Steam : Steam taken directly from a boiler retains the same temperature and pressure as when it was produced and is said to be "saturated". By taking such steam and heating it further, a much greater volume of gas can be produced for a small amount of extra heat. Tests have shown that superheating can result in a 25% economy in coal and 30% in water. Introduced in the 1900's and used mainly on the larger main-line engines. Superheated steam will always be dry.

Superheterodyne Receiver : A radio receiver that combines a locally generated frequency with the carrier frequency to produce a lower-frequency signal (IF, or intermediate frequency) that is easier to demodulate than the original modulated carrier.

Surface Mount Technology : A method of achieving miniaturisation in which electronic components are glued to a special printed circuit board which is then flow soldered to make the connections. The placement of the components may be performed by robot machines. Many recent ECUs have a mixture of surface mount and conventional through board components.

Surface Transfer Impedance : The ratio of the voltage developed across a transmission line to the current flowing on the outer surface of the screen.

Surface Water : Water which is contained in lakes, rivers, and oceans.

Suspended Growth Reactor : A reactor in which the microorganisms are suspended in the wastewater. Examples of suspended growth reactors are activated sludge reactors and anaerobic digesters.

Swallow Counter : The Swallow Counter is one of the three building blocks (swallow counter, main counter, and dual-modulus prescaler) that constitute the programmable divider commonly used in modern frequency synthesisers. The swallow counter is used to control the dual-modulus prescaler which is set to either N or (N+1). At the initial reset state, the prescaler is set to a divide ratio of (N+1), but the swallow counter will change this divide ratio to N when it finishes counting S number of cycles.

Swirl : An orderly rotating motion of the air/fuel charge in the cylinder induced by the incoming flow through the inlet valve and port which may be specially shaped to promote the condition. Useful in the early stages of combustion but usually allowed to then degenerate into random turbulence to spread flame through the charge.

Switch Mode : Uses a switching transistor and inductor to control/regulate the charging voltage/current.

Switched Capacitor Circuit : A circuit methodology, typically implemented in CMOS integrated circuits, that uses clocked switches and capacitors to transfer charge from node to node such that a resistor function is realised. The effective resistance is governed by capacitor size and switching clock frequency.

Switching Regulator : A voltage regulator that uses a switching element to transform the supply into an alternating current, which is then converted to a different voltage using capacitors, inductors, and other elements, then converted back to DC. The circuit includes regulation and filtering components to insure a steady output. Advantages include the ability to generate voltages beyond the input supply range and efficiency; disadvantages include complexity.

Synchronous Rectification : In switch-mode power supplies, the "steering" diode is replaced or paralleled with a FET switch to reduce losses and thereby increase efficiency. The FET is off during the inductor charge cycle, and then turned on as the inductor discharges into the load.

Synergism is the act of working together. Two chemicals which are synergistic have a greater effect together than the sum of their individual effects. The effect can be either positive or negative.

System Timing and Control : Clock generation and distribution systems and components. May include the means for clock control such as spread-spectrum clock generation for EMI reduction, skew rate control, rate dividers, rate control, width, delay, and phase adjustment.

System : An arbitrarily defined area or volume surrounded by a boundary and possessing specific inputs, outputs, and reactions.

T

T (Tera) : The prefix T or tera indicates the unit should be multiplied by 1012

T Flip Flop : A style of flip flop that toggles on a clock edge, with the input high and does not change when the T input is low. A T flip flop can be produced by using a JK flip flop.

T1 : T1 is standard for digital transmission in the United States. It is a digital transmission link with a capacity of 1.544Mbps. T1 uses two pairs of normal twisted wires, the same as found in most residences. T1 normally handles 24 voice conversations, each one digitised at 64kbps. With more advanced digital voice encoding techniques, T1 can handle more voice channels.

T3 : A type of data connection capable of transmitting a digital signal at 44Mbps. T3 lines are often used to link large computer networks, such as those that comprise the Internet.

Tachometer : A small ac or dc generator, sometimes referred to as a rate generator, that converts its shaft speed into an electrical output. The tachometer is frequently used in servo-systems to sense the velocity of a load. An instrument that measures the rate at which a shaft is turning.

Tank Circuit : A tuned circuit used to temporarily store energy. Also referred to as a parallel-resonant circuit.

Taper : In a potentiometer, taper refers to how the resistance varies as the pot's armature is rotated (or, for a slide pot, as its wiper slides; or for a solid state pot like the DS1802, as its input voltage is varied). For a pot with a linear taper, the resistance varies linearly as the wiper moves. For a pot with a logarithmic (log) taper, the resistance varies logarithmically with the wiper's motion. When used in an amplifier circuit, the output varies slowly as the pot is operated at the low end and varies more and more rapidly as the pot is operated toward the high end. This is also called an audio taper because it is most commonly used for audio volume controls.

Tapped Delay Line : A delay line with multiple tapped outputs so the amount of delay may be selected by choosing another tap or point on the delay line. To change the delay on this type of circuit the output needs to be taken from a different output pin.

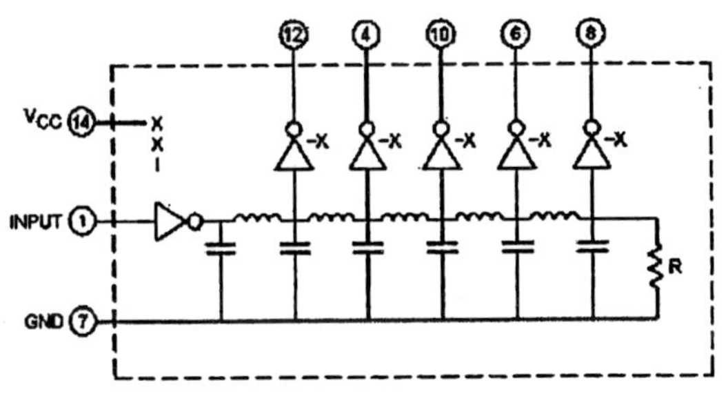

Tapped Delay Line

Tapped Resistor : A wire-wound, fixed resistor having one or more additional terminals along its length, generally for voltage-divider applications.

TCP/IP (Transmission Control Protocol/Internet Protocol) : The protocols or conventions that computers use to communicate over the Internet.

TCXO (Temperature Compensated Crystal Oscillator) : A crystal oscillator that includes circuitry that compensates for temperature variations, to maintain a more constant frequency.

TDD : Time Division Duplex, the second variation of WCDMA especially suited to indoor environments where there is a need for high traffic density.

TDM : Time Division Multiplexing, a scheme in which numerous signals are combined for transmission on a single communications line or channel. Each signal is broken into many segments, each having very short duration.

TDMA (Time Division Multiple Access) : A method of digital wireless-communications transmission. TDMA allows many users to access (in sequence) a single radio-frequency channel without interference, because it allocates unique time slots to each user within each channel.

Teak : A strong and durable wood that doesn't shrink much and is comparatively easy to work.

TEC : A thermoelectric cooler (TEC) is a small cooling device that relies on a Peltier junction. Composed of two conductors made of different materials, a Peltier junction (discovered in 1833 by J.C. Peltier) acts as a heat pump which can cool or warm when current is passed through it. The small size of the TEC allows precision thermal control of individual components such as fiber optic laser drivers, precision voltage references, or any other temperature critical device.

Technical Ceramic : A ceramic that exhibits a high degree of industrial efficiency through carefully designed microstructures and superb dimensional precision.

Tee Connector : A style of connector shaped like a 'T'. The connectors may be any gender in any combination. A BNC style Tee is shown to the right, but a Tee connector could be almost any style. M

Telecommunications : The transmission, emission, or reception of signs, signals, writings, images, or sounds. This is done by visual, oral, wire, radio, or other means. The term Telecommunications may also be abbreviated as Telcom.

Television : A system for transmitting picture and sound over a distance, primarily via the standards for NTSC, PAL, or HDTV.

Temperature : The average kinetic energy of the atoms or molecules of a body or substance, perceived as warmth or coldness. Measured in degrees Fahrenheit, Celsius, or Kelvin.

Temperature Coefficient : The ratio of the change in a parameter to the change in temperature. The amount of change of resistance in a material per unit change in temperature.

Temperature Comparator : An integrated circuit with a digital output that indicates whether a measured temperature is above or below a predetermined threshold.

Temperature Inversion : The condition in which warm air is formed above a layer of cool air that is near the earth's surface.

Temperature Sensor : Temperature sensor that uses an external diode-connected transistor as the sensing element to measure temperatures external to the sensor (for example, on a circuit board or on the die of a CPU). Generally produces a digital output.

Temperature Switch : A circuit that opens and closes a conductive path based on temperature.

Tempering : A heat treatment process where a steel that may have been heat treated to harden it is reheated to modify it's structure

Tempest : A term normally used to describe compromising emanations. These emanations are unintentionally radiated signals that could disclose classified information.

Tensile Strength : The greatest stress a substance can withstand along its length without tearing apart.

Tented Via : A via with dry film solder mask which completely covers both the via pad and its plated-through hole [PTH].

Terminal Board : Also called a terminal strip. An insulating base or slab equipped with terminals for connecting wiring.

Terminal Diagram : A diagram of a switch, relay, terminal board, or other component showing the connections to the equipment.

Terminal Lug : A device attached to a conductor to permit connection to a terminal.

Test Equipment : A general term applied to devices used to test electrical and electronic circuits.

Test Points : Locations in equipment or devices that are accessible to the technician's test probes where operating voltages or signals can be monitored.

Tetrode Tube : A four-electrode electron tube containing a plate, a cathode, a control grid, and a screen grid.

T Flip Flop : A style of flip flop that toggles on a clock edge, with the input high and does not change when the T input is low. A T flip flop can be produced by using a JK flip flop

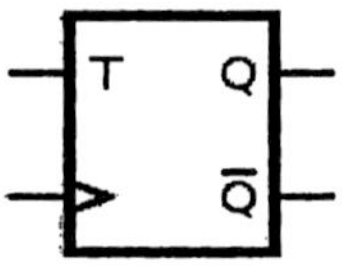

T Flip Flop

THD : ***Total Harmonic Distortion (THD)*** : A measure of signal distortion which assesses the energy that occurs on harmonics of the original signal. It is specified as a percentage of the signal amplitude. As an example, if a 12kHz signal is applied to the input, THD would look at energy on the output occurring at 24kHz, 36kHz, 48kHz, etc. and compare it to the energy occurring at 12kHz.

THD+N : Total Harmonic Distortion Plus Noise (THD+N) is the sum of the two most important distortion components. THD is the distortion that occurs on harmonics of the original signal it is correlated with the signal.

Therm : This is the unit of energy used by the gas industry and 1 therm is equivalent to 100000Btu = 105.5MJ.

Thermal Control Circuit : Circuit to monitor and control the temperature of something. For example the integrated temperature controller in Intel's processors.

Thermal Impedance : The resistance to heat transfer.

Thermal Inertia : The capacity of a soldering iron to generate and maintain a satisfactory soldering temperature while giving up heat to the material being soldered.

Thermal Management : The use of various temperature monitoring devices and cooling methods, such as forced air flow, within a processor or FPGA-based system, to control overall temperature of ICs and internal cabinet temperatures.

Thermal Monitor : The integrated thermal control system used in Intel's processor devices.

Thermal Noise : The white noise generated by a passive component.

Thermal Resistance : A figure of merit for heat transfer [temperature drop divided by power dissipation] between two points in a conductive system.

Thermal Runaway : A conduction that exists when heat causes more electron-hole pairs to be generated, which, in turn, causes more heat and which may eventually cause component destruction [in diodes and transistors]. Several semiconductor characteristics are a function of temperature, power loses increase with increasing temperature causing more and more heat to build within the device until the junction temperature is exceeded.

Thermal Shutdown : Deactivating a circuit when a measured temperature is beyond a predetermined value.

Thermal Trip Element : A circuit breaker trip element that uses the increased bending of a bimetallic strip caused by increased current to open a circuit.

Thermal Via : A via in the Printed Wiring Board [PWB] to effectively transfer heat from the top copper layer of the Printed Circuit Board [PCB] to the inner or bottom copper layers. The thermal vias should make their connection to the internal ground plane with a complete connection around the entire circumference of the plated through hole. Do not use a thermal relief web or spoke connection which impedes the conduction path into the inner copper layer.

Thermal-Magnetic Trip Element : A single circuit breaker trip element that combines the action of a thermal and a magnetic trip element.

THERMDA : Thermal Diode Anode pin on AMD and Intel processors.

THERMDC : Thermal Diode Cathode pin on AMD and Intel processors.

Thermionic Emission : Emission of electrons from a solid body as a result of elevated temperature.

Thermistor : A temperature-dependent resistor with a high temperature coefficient, usually composed of sintered semiconductor material.

Thermochron : A device which measures and records (logs) temperature. Trademark of Dallas Semiconductor.

Thermocline : The depth at which an inflection point occurs in a lake temperature profile.

Thermocouple : A temperature sensor formed by the junction of two dissimilar metals. A thermocouple produces a voltage proportional to the difference in temperature between the hot junction and the lead wire (cold) junction.

Thermostat : Circuit that indicates whether a measured temperature is above or below a particular temperature threshold or trip point. Used for thermal protection and simple temperature control systems.

Thermo-Time Switch : A two terminal bimetallic switch and heating coil device threaded into the coolant system, and using the engine block as a ground terminal, which controls the cold start injection system.

Thermtrip : Pin name of the Thermal Trip digital output on Intel Pentium processors. The pin is asserted at a nominal die temperature of 135 degrees-C.

Thermtripl : Pin name of the thermal trip output pin of AMD processors. The pin is asserted at a nominal die temperature of 125°C.

Thiols **:** Organic compounds which contain the "-SH" functional group. : Also called mercaptans.

Three-Way Catalyst **:** A type of catalytic converter used with stoichiometric mixture precisely controlled by a Lambda sensor. Nitrogen oxide compounds are broken down into the separate constituents of nitrogen and oxygen, the latter then being used to oxidise carbon monoxide and unburnt hydrocarbons.

Throttle Potentiometer. Device attached to throttle spindle or linkage and used by the ECU to detect closed throttle, approach of full throttle, rate of throttle movement, and throttle angle. Racing systems often use the throttle potentiometer to provide the load signal. Produces a voltage output proportional to throttle angle. May also provide a similar signal for use by a transmission controller.

Throttle Switch **:** Device attached to throttle spindle or linkage and used by the ECU to detect certain conditions - usually closed throttle and approach of full throttle. D Jetronic also has sequential contacts to produce extra fuel pulses as throttle opens.

Through-Hole **:** A method for mounting components on a printed circuit board (PCB) in which pins on the component are inserted into holes in the board and soldered in place.

Thyristor **:** A type of high voltage current switching device which, once triggered into conduction by a brief pulse, will remain 'on' until current ceases to flow through it. Widely used as current choppers in a.c. mains power control circuits, (lamp dimmers, speed controllers etc.) but also used in Capacitor Discharge ignition systems.

Timber **:** A brief description of timbers and there uses from Acacia to Walnut.

Time Diversity **:** In radio systems, Time Diversity spreads a signal across multiple channels by placing multiple versions of the signal in different time slots.

Timer **:** A circuit used to track the passage of time. A 555 IC is a common integrated circuit used as a timer via the charging rate of an RC circuit. Also refer to following 555 circuit descriptions.

TINI (Tiny Internet Interface) : The industry's smallest web server, TINI is a microcontroller that includes the facilities necessary to connect to the Internet. The platform is a combination of broad based I/O, a full TCP/IP stack and an extensible Java runtime environment that simplifies development of network-connected equipment.

Titanium : A metallic element resembling iron, characterised by it's strength, lightness and corrosion resistance.

Toluene : A high octane aromatic compound used as a gasoline blend component, particularly for racing gasolines.

Totem Pole : A standard CMOS output structure where a P-channel MOSFET is connected in series with an N-Channel MOSFET and the connection point between the two is the output. The P-FET sits on top of the N-FET like a "totem pole." Both gates are driven by the same signal. When the signal is low, the P-FET is on; when the signal is high, the N-FET is on. This creates a push-pull output using just two transistors.

Trabecular Meshwork : Mesh-like structure inside the eye at the iris-scleral junction of the anterior chamber angle. Filters aqueous fluid and controls its flow into the canal of Schlemm, prior to its leaving the anterior chamber.

Trace Contaminants : Contamination found in trace (very low) levels.

Trace : To monitor the execution of a program and report the sequence of actions carried out.

Transceiver : A device that contains both a transmitter and receiver. Examples: Interface devices (such as line drivers and receivers, RS-232, RS-485, RS-422, CAN, LVDS, or USB).

Transconductance : The gain of a transconductance amplifier (an amp in which the a change in input voltage causes a linear change in output current). The basic gain of vacuum tubes and FETs is expressed as transconductance. It is represented with the symbol gm. The term derives from "transfer conductance" and is measured in siemens (S), where 1 siemens = 1 ampere per volt. It was formerly measured as "mho" (ohm spelled backwards).

Transconductance Amplifier : An amplifier that converts a voltage to a current. Also known by several other terms. One synonym is OTA, or operational transconductance amplifier, a term that marries the terms. transconductance amplifier and operational amplifier.The term derives from "transfer conductance" and is measured in siemens (S), where 1 siemens = 1 ampere per volt. It is represented with the symbol gm. The basic gain of vacuum tubes and FETs is expressed as transconductance transformer

Transducer Electronic Data Sheet : A Transducer Electronic Data Sheet, or TEDS, is a method for plug-and-play sensor and transducer hook-up in which the sensor's calibration information is stored within the device and downloaded to the master controller when requested. A standardised TEDS specification is being developed by the IEEE, as IEEE P 1451.4.

Transducer : General term for a device which converts some physical dimension such as temperature, pressure, or movement, into an electrical signal.

Transfer : Transfer refers to the amount of data transferred across a digital interface, exclusive of any extra bits used to encode the data. The number of data transfers is less than the number of bits transmitted when encoded data has more bits than the raw data. As an example, a PCIe serial bus uses 10 bits to encode eight data bits. (Extra bit space may be used to encode a clock, error-detection redundancy, etc.) Data rates are commonly expressed in transfers per second, gigatransfers per second (GT/s) and megatransfers per second (MT/s).

Transformer : An inductive electrical device for changing the voltage of alternating current. A transformer consists of two magnetically coupled coils. Alternating current in one (called the "primary") creates a changing magnetic field which induces a current in the second coil (the "secondary"). A core made of iron or ferrite generally connects the two coils, but higher frequency devices can work without a ferrous core.

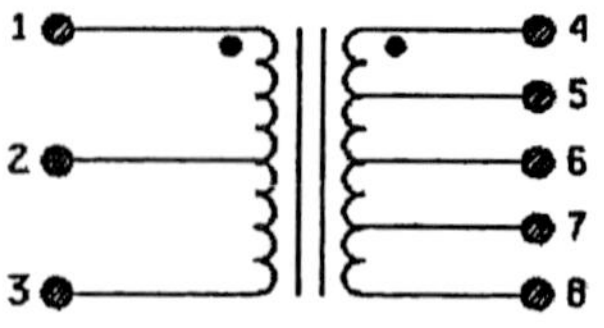

Transformer Symbol

Transient Intermodulation Distortion : Transient intermodulation distortion, or TIM, occurs in amplifiers that employ negative feedback when signal delays make the amplifier incapable of correcting distortion when exposed to fast, transient signals.

Transimpedance Amplifier : An amplifier which converts a current to a voltage. It is a familiar component in fiber-communications modules.The unit for trans resistance is the ohm.

Transistor : A basic solid-state control device which allows or disallows current flow between two terminals, based on the voltage or current delivered to a third terminal. Usually built from silicon but can be constructed from other semiconductor materials. There are two major types: The FET (field-effect transistor) and the bipolar junction transistor (BJT). The first transistor was invented in 1947 at Bell Labs by Michael John Bardeen, Walter Brattain and William Shockley.

Transition Delay : Fault characterised by a particular gate or gate connection being too slow-to-rise or too slow-to-fall to meet the overall timing requirements of a circuit.

Transmission Line : A pair of conductors separated by a dielectric.

Transmitter : A circuit that accepts signals or data in and translates them into a form that can be sent across a medium (transmitted), usually over a distance. The medium can be wireless or wired. Exmples: A radio transmitter that modulates the signal on a carrier and transmits it over the airwaves. An ultrasonic transducer that sends the signal over ultrasound frequencies.

Transpiration **:** The loss of water from plants through leaves and other parts. This loss can be a significant amount of water during very dry periods.

Transverse Mode **:** A propagation mode in which the two conductor of a pair carry equal and opposite currents.

Trickling Filter **:** An attached growth biological process in which the microbial film is attached to non-moving rock or plastic media.

Trigger Switch **:** A switch, usually paired, which, by briefly completing a circuit to ground, triggers the injection functions of the Bosch D Jetronic system. May take the form of simple contact breakers or sealed magnetically operated switches.

Troposphere **:** The lower atmosphere, from the earth's surface to approximately 12 km. This portion of the earth's atmosphere contains about 95 percent of the atmospheric gases. The temperature gradually declines through this region.

TS 16949 **:** TS 16949 is an ISO Technical Specification that aligns previous American, German (VDA6.1), French (EAQF) and Italian (AVSQ) automotive quality systems standards within the global automotive industry. Together with ISO 9001:2000, ISO/TS 16949:2002 specifies the quality system requirements for the design/development, production, installation and servicing of automotive related products.

TVS (Transient Voltage Suppressor) **:** Semiconductor device designed to protect a circuit from voltage and current transients. Typically implemented as a large silicon diode operating in avalanche mode to absorb large currents quickly.

Twisted pair **:** A transmission line in which two insulated conductors are twisted together.

Two-Stroke Cycle : Poor fuel economy is made up for by the reduction in components over the four-stroke engine and specific power per swept capacity. Making the two-stroke particularly popular for aircraft applications and models in general.

Two-Wire Open Line : A parallel line consisting of two wires that are generally spaced from 2 to 6 inches apart by insulating spacers.

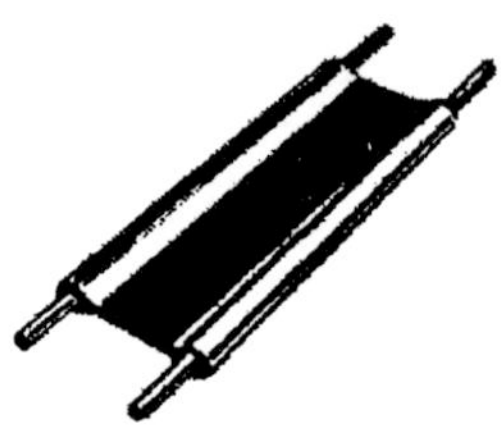

Twin Lead Cable

Two-Wire Ribbon (Twin Lead) : A parallel two-wire line in which uniform spacing is assured by two wires embedded in a low-loss dielectric. Design note; Replace any twin-lead cable with a coaxial cable; a coaxial cable offers the best protection from all types of outside interference.

U

UART (Universal Asynchronous Receiver-Transmitter) : An IC that converts parallel data to serial, for transmission; and converts received serial data to parallel data.

UDL (User Defined Logic) : Special purpose logic added by the SOC chip integrator (e.g. not a purchased IP), for use as "glue" logic, or part of the system chip product differentiated.

UI : Unit interval (used to describe jitter generation); user information; user interface.

UJT : Uni-Junction Transistor. A three-terminal semiconductor device having one junction and a stable negative-resistance characteristic over a wide temperature range.

U-Law : A companding algorithm used in digital communications to reduce the dynamic range of an audio signal.

ULP (Upper Level Protocol) : The protocol which runs on top of Fibre Channel through the FC-4 layer. Typical ULPs running over Fibre Channel are Small Computer System Interface (SCSI), Internet Protocol (IP), High Performance Parallel Interface (HIPPI), and Intelligent Peripheral Interface (IPI).

Ultimate Biochemical Oxygen Demand (BODu) : The total amount of oxygen required to oxidise any organic matter present in a water, i.e. after an extended period, such as 20 or 30 days.

Ultimate Disposal : The process of returning residuals back to the environment in a form which will have the minimal or reduced negative environmental impacts.

Ultrahigh Frequency : The band of frequencies from 300 megahertz to 3 gigahertz.

UMTS (Universal Mobile Telecommunications System) : A future mobile communications system which, among other features, will offer direct connection between terminals and satellites.

Unbalanced Crystal Mixer : A circuit consisting of a section of coaxial transmission line one-half wavelength long that is tuned to the difference (intermediate) frequency between the local oscillator and RF echo signals.

Unbalanced Line : A transmission line, such as a coaxial cable, in which the magnitudes of the voltages on the two conductors are not equal with respect to ground.

Unbalanced Pair : A transmission line in which the voltages on the two conductors are unequal with respect to earth.

Uncertainty : An expression of the combined errors in a test measurement process.

uniform Resource Locator (URL) : The name (or string of characters) that uniquely identifies each Web site. For example, www. tmworld.com.

Unijunction Transistor (UJT) : A three-terminal, semiconductor device with a negative resistance characteristic that is used in switching circuits, oscillators, and wave-shaping circuits. The leads are Anode, Gate, and Cathode.

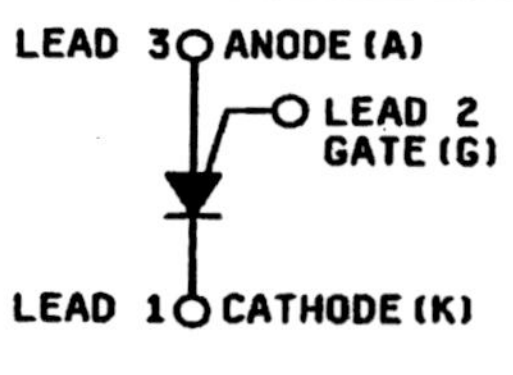

Unijunction Transistor

Uninterruptible Power Supply : An uninterruptible power supply (UPS) is a device that maintains power in the event of a failure. A UPS commonly includes a battery that is kept charged and ready. When power fails, the battery supplies power, as long as it lasts. When the battery fails, a UPS may contain circuitry that triggers an orderly shutdown.

UniqueWare Serialised : A factory-programming service for 1-Wire EPROM chips with customer-specified data. Service provides one serialisation file for customers to create identifiers in silicon.

Unit Size : The standards adopted to make comparisons between things of like value (for example, the unit size for conductors is the mil-Foot).

Unit : An assembly or any combination of parts, subassemblies, and assemblies mounted together. Normally capable of independent operation. A single object or thing.

Universal Time Constant Chart : A chart used to find the time constant of a circuit if the impressed voltage and the values of R and C or R and L are known.

Untapped Delay Line : A simple fixed circuit delay with out adjustment or programmability.

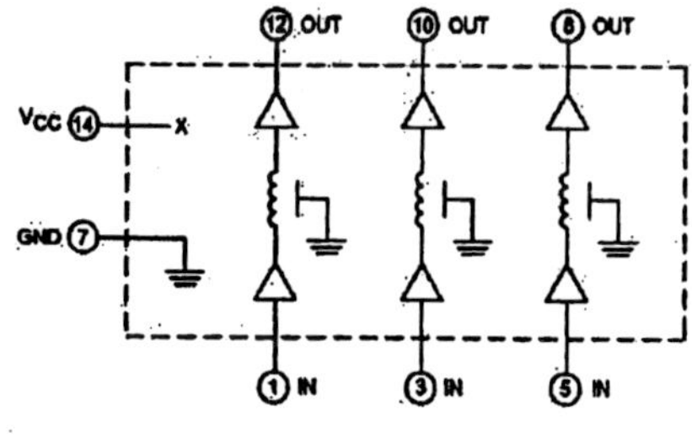

Untapped Delay Line

Untuned Line : Another name for the flat or non-resonant line.

Up Link : The frequency used to transmit a signal from earth to a satellite.

Upconverters : A device which provides frequency conversion to a higher frequency, e.g., in digital broadcast-satellite applications.

Upper Sideband : All of the sum frequencies above the carrier.

Upper-Frequency Cutoff : The highest frequency a circuit can pass.

USB (Universal Serial Bus (USB) : A standard port that enables you to connect external devices (such as digital cameras, scanners, keyboards, and mice) to computers. The USB standard supports data transfer at three rates: low speed (1.5MBps), full speed (12Mbps) and high speed (480 MBps).

Utopia Bus : A Bus Standard used as the Test and Operations Physical Interface for ATM [Asynchronous Transfer Mode] standard.

Uvea, Uveal Tract : Pigmented layers of the eye (iris, ciliary body, choroid) that contain most of the intraocular blood vessels.

Uveitis : Inflammation of the uveal tract. It may be anterior involving the iris and ciliary body (iridocyclitis) or posterior involving the choroid (choroiditis).

UWB : Ultra-Wideband (UWB) is a communications technology that employs a wide bandwidth (typically defined as greater than 20% of the center frequency or 500MHz). UWB is usually used in short-range wireless applications but can be sent over wires. Ultra-Wideband advantages are that it can carry high data rates with low power and little interference.

V

Vacancy **:** A normally occupied lattice site from which an atom or ion is missing.

Vacuum Evaporation **:** Process of producing thin film components.

Vacuum Tube **:** An electron tube evacuated such that its electrical characteristics are essentially unaffected by the presence of residual gas.

Vacuum Tube

Validation **:** A "post-silicon" process to prove (with evidence) a design to be valid. For our purposes, validation is any use of special purpose test hardware to prove the product meets the design intent; as opposed to other usage of the same equipment as a manufacturing screen.

Van Der Waals Bond **:** A secondary interatomic bond between adjacent molecular dipoles, which may be permanent or induced.

Vane Attenuator : A wave guide component used to provide attenuation.

Var Hour : The reactive power, Volt-Ampere used within an hour; as measured by a VAR Hour meter.

Varactor Fm Modulator : An FM modulator that uses a voltage-variable capacitor (varactor).

Varactor : A PN junction semiconductor, designed for microwave frequencies, in which the capacitance varies with the applied voltage. 'Varicap' or 'VVC', Voltage [dependent] variable capacitor [varies in picofarads].

Variable : A representative symbol that can assume any of a given set of values.

Variable Attenuator : An attenuator for reducing the strength of an ac signal either continuously or in steps, without causing signal distortion.

Variable Capacitor : An adjustable Capacitor. In this example the dielectric is air, other dielectrics are possible and additional package styles.

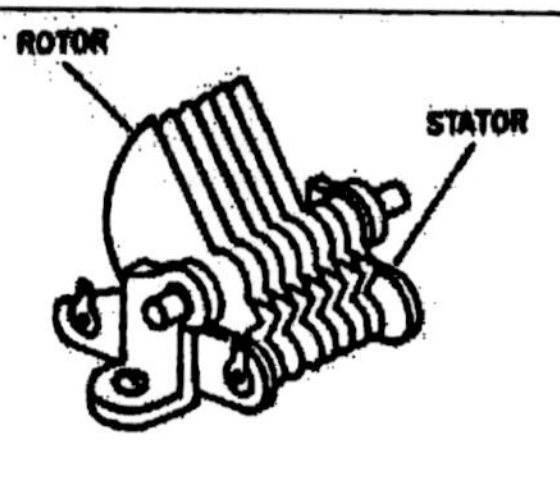

Variable Capacitor

Variable Resistor : A wire-wound or composition resistor, the value of which may be changed over a designed range.

Variable Valve Event System : A general term applied to cover any method of changing some feature of the valve operating cycle whilst the engine is running. Cam advance or retard, changes of lift and duration, and alteration to the number of valves in use, all fall in this category. Can be used to control exhaust emissions, widen torque band, or improve idle quality.

Variac : An autotransformer that contains a toroidal winding and a rotating carbon brush so that the output voltage is continuously adjustable from zero to line voltage

Varistor : A resistor that changes value with applied voltage. A varistor may also be called a VDR [Voltage Dependent Resistor]. Varistors will have a negative voltage coefficient.

Varnished Cambric : Cotton cloth coated with insulation varnish. An insulation used on high-voltage conductors.

VCC : The supply voltage for a circuit is often given as V plus a double-letter suffix. The double letter is usually related to the lead of the transistors that are commonly connected to that supply or to a resistor that connects to that supply. Examples: VCC is a positive-voltage supply and the collector terminal of bipolar transistors is connected to the VCC supply or to a load which connects to VCC.

VCD *(Verilog Change Dump)* : Output from Verilog used for pattern generation. Typically this output has to be translated into a format that is readable by a tester. This translation process is typically performed by several third party and in-house tools.

VCO (Voltage-Controlled Oscillator) : An oscillator device in which output frequency is proportional to its input voltage.

VCTCXO (Voltage Controlled, Temperature Compensated Crystal Oscillator) : A TCXO which offers the ability to control the oscillation frequency with an analog voltage.

VCXO (Voltage Controlled Crystal Oscillator) : An oscillator that uses a crystal to establish its frequency but will vary its frequency as an analog control voltage varies.

VD : Voltage DMOS. In a power supply circuit, voltage supplied by a MOS circuit, measured next to the MOS device.

VDD : In a power supply circuit supplied by a MOS circuit, the voltage supplied by a MOS device but referenced next to the power supply and not next to the DMOS device.

VDSL (Very High Data-Rate Digital Subscriber Line) : A method for delivering high-speed digital services on the standard twisted pair used for voice phone lines. VDSH operates at data rates from 12.9Mbps to 52.8Mbps.

Vector : A parallel vector is the functional logic applied to a DUT during one clock cycle.

Velocity Ratio : The ratio of propagation velocity to the speed of light.

Verification : A "pre-silicon" process done during development, for gaining confidence, that a Design will produce a pre-defined result. An output of verification may be translated into ATE vectors (used for functional testing).

Vertical Sync : The portion of a video signal that indicates the end of a field of video information. This sync pulse is used by video equipment to maintain field synchronisation with the incoming video signal.

Vicker's Hardness : Determined by indenting the material with a diamond pyramid under a specific load and then measuring the size of the impression.

Video Encoder : A device that converts RGB video to composite video.

Viscoelasticity : A type of deformation exhibiting the mechanical characteristics of viscous flow and elastic deformation.

Viscosity : The ratio of the magnitude of an applied shear stress to the velocity gradient that it produces.

Visual Acuity : Assessment of the eye's ability to distinguish object details and shape, using the smallest identifiable object that can be seen at a specified distance (usually 20 ft. or 16 in.).

Visual Field : Full extent of the area visible to an eye that is fixating straight ahead.

Vitreous : The transparent gelatinous material which fills the interior of the eye behind the lens. It allows the eye to maintain its shape.

VLSI (Very Large-scale Integration) : VLSI refers to an IC or technology with many devices on one chip. The question, of course, is how one defines "many." The term originated in the 1970s along with "SSI" (small-scale integration), "LSI" (large-scale), and several others, defined by the number of transistors or gates per IC. The terms LSI and VLSI are now usually used as general terms, referring to a product or technology that subjectively has more devices than typical products in the category.

Volt : Unit of measure for electromotive force (EMF), the electrical potential between two points. An electrical potential of 1 volt will push 1 ampere of current through a 1-ohm resistive load. Using a common plumbing analogy, voltage is similar to water pressure and current is analogous to flow (e.g. liters per minute).

Voltage Doubler : A capacitor charge pump circuit which produces an output voltage which is twice the input voltage.

Voltage Identification Digital : Voltage Identification Digital, or VID, is a circuit concept developed to provide the central processing unit (CPU) of a computer with the appropriate supply voltage. Instead of having a power supply unit generate some fixed voltage, the CPU uses a small set of digital signals, the VID lines, to instruct an on-board power converter of the desired voltage level.

Voltage Regulator : A circuit which is connected between the power source and a load, which provides a constant voltage despite variations in input voltage or output load.

Voltage-to-frequency Converter (VFC) : A device that converts an analog input voltage into a sequence of digital pulses with a frequency that is proportional to the input voltage.

Volt-Ampere : A volt-ampere (VA) is the voltage times the current feeding an electrical load. A kilovolt-ampere (kVA) is 1000 volt-amperes. Electrical power is measured in watts (W): The voltage times the current measured each instant. In a direct current system or for resistive loads, the wattage and VA measurements will be identical. But for reactive loads, the voltage and current are out of phase and the volt-ampere spec will be greater than the wattage. For determining power, watts are appropriate. For determining capacity for the driving circuits (circuit breakers, wiring, and uninterruptible power supplies, for instance), VA is appropriate.

Volumetric Efficiency : The components of the air intake system restrict the amount of air that can be drawn into an engine. It can only be used with four-stroke engines as they have a distinct induction process. It is defined as the volume flow rate of air into the intake system divided by the rate at which volume is displaced by the piston.

Vortex Veins : Four veins (2 superior, 2 inferior) that provide the main blood outflow from the eyes, exiting near the equator on each side of the superior and inferior recti and draining into the superior and inferior ophthalmic veins

VSWR (Voltage Standing Wave Ratio) : VSWR is a measure of how efficiently radio-frequency power is transmitted from a power source, through a transmission line, into a load (for example, from a power amplifier through a transmission line, to an antenna). In an ideal system, 100% of the energy is transmitted.

Vulcanisation : Nonreversible chemical reaction involving sulfur or other suitable agent wherein cross-links are formed between molecular chains in rubber materials.

W

W **:** Watt (W) is the unit for measuring power. In physics terms, one watt is one Joule of energy transferred or dissipated in one second. Electrical power is calculated as: Watts = Volts x Amps x Power Factor. The power factor can be disregarded for DC circuits or for AC circuits with a resistive load (it is 1 in those situations).

Wafer **:** Semiconductor manufacturing begins with a thin disk of semiconductor material, called a "wafer." A series of processes defines transistors and other structures, interconnected by conductors to build the desired circuit. The wafer is then sliced into "dice" which are mounted in packages, creating the IC.

Wafer Fab **:** Semiconductor processing facility which turns wafers into integrated circuits. A typical wafer fab employs a series of complex steps to define conductors, transistors, resistors, and other electronic components on the the semiconductor wafer. Imaging steps define what areas will be affected by subsequent physical and chemical processes.

Wafer Map **:** A plot of the viable dice on a wafer showing pass/fail information, parameter variation, or some other characteristic.

Wafer Switch **:** A rotary switch in which the contacts are arranged on levels. Each level (wafer) is electrically independent but mechanically connected by the shaft of the switch. Rotary Switch vendors.

Warm Up Enrichment **:** Extra fuel provided by the ECU according to engine temperature as sensed by the coolant temperature sensor. The overall fueling is proportionately enriched and acceleration, cranking, and other additional enrichment factors are usually enhanced as necessary.

Waste Minimisation : The elimination or reduction of a waste prior to its generation. This is accomplished by process changes rather than waste treatment methods.

Wasted Spark Ignition : A distributorless arrangement where spark plugs fire twice per cycle, the second spark being 'wasted' because it occurs while exhaust is being scavenged. Has the advantage of simplified triggering arrangements usually used with double ended coils firing two cylinders simultaneously.

Wastewater : Consumed or used water from a municipality or industry that contains dissolved and/or suspended matter.

Watchdog : A feature of a microprocessor supervisory circuit that monitors software execution in a microprocessor or microcontroller. It takes appropriate action (assert a reset or nonmaskable interrupt) if the processor gets stuck in an infinite execution loop.

WatchDog Timer : A timer used by a cpu that will cause the CPU to reset if the CPU does not service [signal] the watchdog time at certain intervals. If the Watch Dog Timer does not receive a signal within a certain length of time it will cause the processor to reset itself.

Watt : The unit of electrical power that is the product of voltage and current.

Wattage Rating : A rating expressing the maximum power that a device can safely handle.

Watt-Hour Meter : A meter used to measure electrical energy.

Watt-Hour : A practical unit of electrical energy equal to one watt of power for one hour.

Wattmeter : A meter used to measure electrical power.

Wave Motion : A recurring disturbance advancing through space with or without the use of a physical medium.

Wave Train : A continuous series of waves with the same amplitude and wavelength.

Wave Winding : An armature winding in which the two ends of each coil are connected to commutator segments separated by the distance between poles.

Waveform Analysis : Observation displays of voltage and current variations with respect to time or by harmonic analysis of complex signals.

Waveform : The shape of the wave obtained when instantaneous values of an ac quantity are plotted against time in rectangular coordinates.

Wavefront : A small section of an expanding sphere of electromagnetic radiation that is perpendicular to the direction of travel of the energy.

Wave Guide Duplexer : TR and atr tubes housed in a resonant cavity attached to a wave guide system.

Wave Guide Mode of Operation : Any particular field configuration in a wave guide that satisfies the boundary conditions. Usually divided into two broad types: the transverse electric (TE) and the transverse magnetic (TM) modes.

Wave Guide Post : A rod of conductive material used as impedance changing devices in waveguides.

Wave Guide Screw : A screw that projects into a wave guide for the purpose of changing the impedance.

Wave Guide : A rectangular, circular, or elliptical metal pipe designed to transport electromagnetic waves through its interior.

Wavelength : The distance, usually expressed in meters, travelled by a wave during the time interval of one complete cycle. It is equal to the velocity divided by the frequency.

Wavemeters : Calibrated resonant circuits that are used to measure frequency. An instrument for measuring the wavelength of an RF wave.

Weak Acid : An acid that does not ionise completely under the conditions of interest. : Examples include acetic acid, carbonic acid, and hypochlorous acid.

Wetland : Semi-aquatic land, that is land that is either inundated or saturated by water for varying periods of time during each year, and that supports aquatic vegetation which is specifically adapted for saturated soil conditions.

Wheatstone Bridge : An ac bridge circuit used to measure unknown values of resistance, inductance, or capacitance. The resistor labeled Rx in the diagram is the component to be measured.

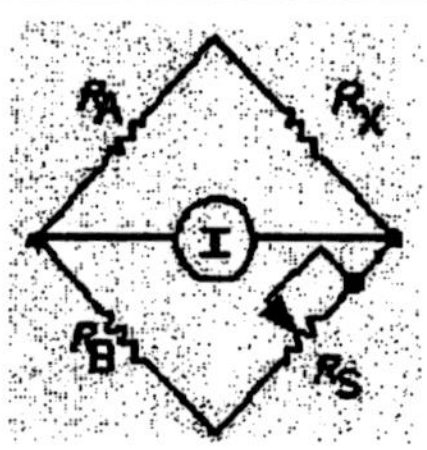

Wheatstone Bridge

White Metal Modelling : So what is white metal and why is it so good for modelling especially unusual subjects. Well after the cost of material and labour the moulding process is relatively cheap. Therefore short runs of kits can be produced and of a variety of different subjects can be kitted.

White Metal : Used to describe low melting point metal. It is used to describe any metal except copper and gold.

Wideband : A classification of the information capacity or bandwidth of a communication channel. Wideband is generally understood to mean a bandwidth between 64kbits/s and 2Mbit/s.

WiMax : WiMax (Worldwide Interoperability for Microwave Access) is a "last mile," broadband, wireless access mechanism which c potentially replace DSL and Cable Modem. Defined by the IEEE 802.16 standards. While Wi-Fi covers a small area with a radius of a few hundred meters, WiMax can cover up to 6 miles with only one base station. Also known as WirelessMAN, for "Wireless Metropolitan Area Networks."

Window Comparator : A device, usually consisting of a pair of voltage comparators, in which output indicates whether the measured signal is within the voltage range bounded by two different thresholds.

Window Watchdog : A special subset of the watchdog timer feature found on microprocessor supervisory circuits. It is used to monitor software execution and assert a reset or an NMI if the processor gets stuck in a loop. This feature not only looks for periodic transitions on its input within a preprogrammed timeout period, but it also looks to see if there are "too many" transitions within its timeout period.

Wireless : Radio-frequency devices, circuits, or communications methods.

Wire Wrapping : A method of attaching a wire between components with out using traces on a Printed Wiring Board [PWB]. The components are placed in a socket or pin which have a wire wrap terminal, while the board it self is blank with either holes or plated-through holes to accept the pins. The components are then interconnected by twisting wires around the wire wrap pins attached to the components.

Wire Wrap Terminal

Wireless Sensor Network : Wireless Sensor Network, or WSN, is a network of RF transceivers, sensors, machine controllers, microcontrollers, and user interface devices with at least two nodes communicating by means of wireless transmissions.

WLBI (Wafer Level Burn In) : Wafer Level Burn In is a manufacturing technique where a particular product is burned in as full wafers to take advantage of massive parallelism and higher temperature acceleration factors.

WLL (Wireless Local Loop) : Any method of using wireless communication in place of a wired connection to provide subscribers with standard telephone service.

Wood Screws : The diameter and pilot drill size versus wood screw gauge.

Wrought Iron : Iron containing 1 to 3% slag giving it a characteristic grain. Other properties are more rust resistant than steel and more easily welded.

Wrought Iron : Iron containing 1 to 3% slag giving it a characteristic grain. Other properties are more rust resistant than steel and more easily welded.

WTA (Wireless Telephony Application) : A collection of telephony-specific extensions for call- and feature-control mechanisms that make advanced mobile network services available to end users. WTA essentially merges the features and services of data networks with the services of voice networks.

X

X-Axis : The horizontal axis of a chart or graph. In a gyro, the spin axis of the gyro.

X-Band : A radio frequency band from 5200MHz to 11,000MHz.

XCO (Crystal Clock Oscillator) : An oscillator that relies on a crystal for its frequency reference. A piezoelectric crystal oscillates at a very stable frequency.

Xenon Arc Lamp : A high intensity light that excites xenon gas between two electrodes.

Xlr Connector : A shielded 3-conductor microphone plug.

XNOR : Short for Exclusive-NOR, refer to that term. A logic function and a logic gate.

XOFF : An abbreviation for the ASCII transmission-control character meaning "Transmitter off."

XON : An abbreviation for the ASCII transmission-control character meaning "Transmitter on."

X-Ray Emission : Penetrating radiation similar to light, but with shorter wavelength, that can penetrate human tissue.

Y

Y Axis of Fick : Imaginary line through an eye's center of rotation, connecting the geometric center of the cornea (anterior pole) with geometric center of the back of the eye (posterior pole). Torsional (tilting) eye rotations occur around the axis.

Y/C : Y, C, YUV, Y-Pb-Pr, YCbCr, and Y/C (also known as S-video) are terms that refer to video signal components. The black and white (luminance) portion of the video signal is the "Y" component which, when combined with color components, form a complete picture. The different nomenclatures for the color components reflect different color encoding schemes.

Yagi Antenna : A combination of dipoles to increase the gain.

Y-Amplifier : The vertical amplifier of an oscilloscope.

YAW : Side to side motion in a vehicle. The rotation of an object in a plane around a vertical axis.

Y-AXIS : In a gyro, an axis through the center of gravity and perpendicular to the spin axis. The vertical axis on a chart.

Yield Strength : The stress required to produce a very slight yet specified amount of plastic strain.

YIG : Yttrium-iron-garnet (YIG) is a ferrimagnetic material used for solid-state lasers and for microwave and optical communications devices.

Yoke : A set of coils placed over the neck of a Cathode Ray Tube CRT to move the electron beam position as it points at the face of the CRT.

Youngs Modulus **:** This is the modulus of elasticity applicable to the bending of a beam or stretching of a thin rod. It is defined as the ratio of *stress* over *strain*. Units N/m2. A hard white metal with a bluish tinge. Zinc base alloys are used extensively for diecasting.

Y-Tilt **:** In aerial photography, the deviation of a camera from the vertical caused by the pitch of the aircraft.

Z

Z : Symbol used to indicate impedance.

Z Axis of Fick : Imaginary line that passes vertically through the center of each eye; associated with horizontal eye rotations.

Z-Axis : In a gyro, an axis through the center of gravity and mutually perpendicular to both the X (spin) and Y axes.

Zener Diode : A type of diode which, at a certain voltage (called the Zener voltage), conducts in the reverse direction. Zener diodes rated from 2.4 volts to hundreds of volts are available and can be used to set or limit the maximum voltage in an electrical system. Accuracy varies with current and temperature.

Zener Effect : A reverse breakdown effect in diodes in which breakdown occurs at reverse voltages below 5 volts. The presence of a high energy field at the junction of a semiconductor produces the breakdown.

Zero Adjustment : The process of Nulling out a circuit or system. Setting a circuit to a zero output condition.

Zero Insertion Force Socket : A ZIF socket requires no force to install an IC into the socket. A spring actuated level is used to clamp the device in place once it has been installed.

Zero Vergence : Refers to light rays that are parallel and have no converging or diverging optical power.

ZERO-Bit Insertion : A transmission process when additional "zeros" are inserted after a fixed number of "ones" in a transmission sequence to ensure that data will not be misinterpreted as controlling information.

Zerofill : To fill unused storage locations with the representation of the character denoting "0".

Zeroing : The process of adjusting a synchro to its electrical zero position.

ZIF (Zero Insertion Force) : A class of IC sockets which clamp the IC pins (via a small lever on the side of the socket) after insertion, and thus require no downward force on the IC or its pins to insert it into the socket. Especially useful in applications in which repeated insertions subjects the IC or the socket to wear and breakage.

Zig Zag Package : A component package style that has alternating terminals which are off-set from one another.

ZIGBEE : A standard for short-distance, low-data-rate communications using the frequencies and physical and data layers of the IEEE 802.15.4 PHY specification. Created and maintained by the ZIGBEE Alliance Group.

Zone of Mutual Visibility : The area where the satellite can be seen by both the up- and down-link earth terminals.

Zone of Silence : Skip zone. The region between alternate reflections of a radio wave, in which no signal is detectable.

Zone Plate : A flat glass plate with concentric opaque rings that forms an image caused by diffraction rather than reflection.

Zonules : Radial fibres that suspend the lens from the ciliary body and hold it in position.

Zoom Lens : A lens having internal movable elements that allow the focal length to be changed without changing the image plane.

Zoom : To change the magnification and focal length of a zoom lens.